PEYOTE The Divine Cactus

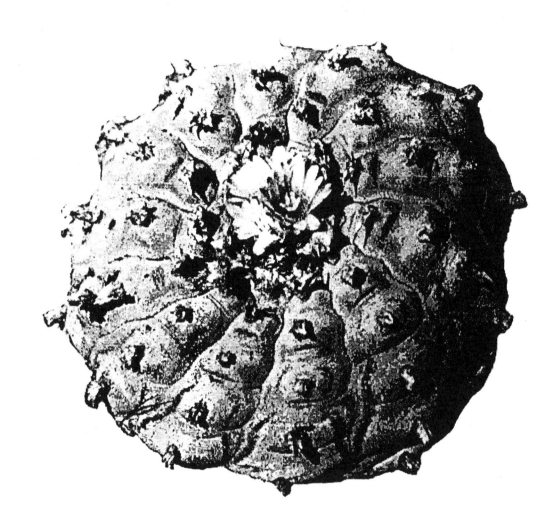

PEYOTE

The Divine Cactus

SECOND EDITION

Edward F. Anderson

THE UNIVERSITY OF ARIZONA PRESS *Tucson*

The University of Arizona Press
Copyright © 1996
The Arizona Board of Regents
All Rights Reserved
First edition published 1980.
Second edition 1996.

♾ This book is printed on acid-free, archival-quality paper.
Manufactured in the United States of America
01 00 99 98 97 96
6 5 4 3 2 1

Library of Congress Cataloging-in-Publication Data
Anderson, Edward F., 1932–
Peyote : the divine cactus / Edward F. Anderson.
— 2nd ed.
p. cm.
Includes bibliographical references and index.
ISBN 0-8165-1653-7 (cloth : acid-free paper). —
ISBN 0-8165-1654-5 (pbk. : acid-free paper)
1. Peyotism.   2. Indians of North America—Religion.
3. Indians of North America—Rites and ceremonies.
4. Peyote.   I. Title.
E98.R3A5   1996
299'.7—dc20   95-50224
CIP

British Cataloguing-in-Publication Data
A catalogue record for this book is available from the British Library.

Publication of this book is made possible in part by the proceeds of a permanent endowment created with the assistance of a Challenge Grant from the National Endowment for the Humanities, a federal agency.

TO ADELE

Contents

List of Illustrations, *xi*

A Word from the Author, *xiii*

Introduction, *xvii*

1. Peyote in Mexico, 3
    The Spanish Conquest and Peyote, 4
    Native American Use of Peyote, 11

2. Peyote in the United States, 25
    Native American Traditions, 25
    Possible Origins of Peyotism, 31
    Native American Oppression, 34
    The Attraction of Peyotism, 39
    The Spread of Peyotism, 42

3. Peyote Ceremonies, 49
    Preparations for the Ceremony, 50
    The Basic Plains Ceremony, 54
    Variations of Peyote Ceremonies, 58
    Visions, 62
    Christian Influences, 62
    Peyote Music, 65
    Memories of a Navajo V-Way Ceremony, 66

4. The Peyote User's Experience, 79
    Phases of the Experience, 83
    Important Aspects of the Experience, 84
    Visions, 92
    Adverse Reactions, 102

5. Medicinal Use of Peyote, 106
    Use by Native Americans, 106
    Early Uses by Anglo-Americans, 111
    Possible Antibiotic Actions, 114
    Psychiatric Uses, 115

6. Pharmacology of Peyote, 120
    Physiological Effects, 120
    Physiological Action of Mescaline, 123
    Dosages, 125
    Toxicity of Mescaline, 125
    Tolerance, 127
    Cytogenetic Effects, 128
    Mescaline, LSD, Psilocybin, and Marijuana, 129

7. Chemistry of Peyote, 133
    Alkaloids, 133
    Peyote Alkaloids, 138
    Peyote Alkaloids in Other Cacti, 141
    Chemically Different Species of Peyote, 143
    Mescaline, 144
    Structure-Activity Relationship of Mescaline, 147
    Biosynthesis of Peyote Alkaloids, 149

8. Botany of Peyote, 153
    Botanical History, 155
    Common Names of Lophophora, *159*
    Plants Confused with Peyote, 162
    Morphology, 165
    Biogeography, 168
    Ecology, 171
    Characteristics, 175
    Evolution, 177
    Cultivation, 178
    Conservation, 180

9. Legal Aspects of Peyote, 183
    Narcotics, Drugs, and Addiction, 185
    Religious Freedom, 187
    *The Comprehensive Drug Abuse Prevention and
        Control Act of 1970, 197*
    The American Indian Religious Freedom Act of 1978, 201
    The United States Supreme Court Decision of 1990, 201
    Congressional Actions Since 1990, 204

Epilogue: The Divine Cactus, 206

Appendix A.
    Peyote Systematics, 209

Appendix B.
    Peyote Alkaloids, 220

Appendix C.
    Excerpts of Current Federal Laws and
    Regulations Pertaining to Peyote, 229
        The Comprehensive Drug Abuse Prevention and
            Control Act of 1970, 229
        The American Indian Religious Freedom Act of
            1978, 233
        The Religious Freedom Restoration Act of 1993, 234
        American Indian Religious Freedom Act
            Amendments of 1994, 235

References Cited, 237

Index, 263

Illustrations

Frontispiece Peyote in flower
(*Lophophora williamsii* [Lemaire] J. Coulter)
1.1 Map: Four peyote-using groups of Native Americans in Mexico, 12
3.1 Peyote "buttons," 51
3.2 Floor plan of the basic Plains peyote ceremony, 54
3.3 Arrangement of the hogan for a Navajo V-Way ceremony, 68
6.1 Chemical structures of LSD and psilocybin, 130
7.1 Anhalinine, 134
7.2 Mescaline, 134
7.3 Δ^9-tetrahydrocannabinol (THC), 137
7.4 Myristicin (3-methoxy-4,5-methylenedioxy), 137
7.5 The basic indole ring, 137
7.6 Chemical structure of a phenol, 139
7.7 Toluene (a non-phenolic compound), 139
7.8 β-phenylethylamine (mescaline), 140
7.9 Reaction of mescaline and formaldehyde to produce anhalinine, 140
7.10 Structure of mescaline, 148
7.11 Epinephrine, 148
7.12 Amphetamine, 148
7.13 3,4,5-trimethoxyamphetamine, 148
7.14 Tyrosine, 151
7.15 Probable biosynthetic pathways of some peyote alkaloids, 151
8.1 First illustration of peyote from *Curtis's Botanical Magazine* of 1847, 156
8.2 Map: Natural distribution of *Lophophora williamsii* and *L. diffusa*, 158
8.3 Habitat of *Lophophora williamsii* in San Luis Potosí, Mexico, 169
8.4 Clump of *Lophophora williamsii* in San Luis Potosí, 170
8.5 Habitat of *Lophophora diffusa* in Querétaro, Mexico, 171
8.6 Cluster of *Lophophora diffusa* growing in Querétaro, 172
A.1 *Lophophora williamsii* (Lemaire) J. Coulter, 211
A.2 *Lophophora diffusa* (Croizat) H. Brav.-Holl., 217

A Word from the Author

Largely because of its special ability to alter the human state of consciousness, peyote has held a place in history for many centuries. An upsurge of interest in psychoactive substances in the 1960s made peyote newly familiar to many. Since that time, its name has often appeared in connection with several social, legal, and scientific issues.

What is it in peyote that causes such unusual effects? Can modern medical science learn anything from Native Americans' use of peyote in curing a wide variety of ailments? What is the Native American Church, and how do its members use peyote? Does anyone have the legal right to use drugs or controlled substances in religious ceremonies?

I had not even heard of peyote in 1957, when Lyman Benson of Pomona College and the Claremont Graduate School notified me that Gordon A. Alles, a research biochemist from Pasadena, California, was looking for a graduate student in botany to serve as a research assistant. Dr. Alles, who gained fame and respect for his pioneering work with amphetamines, had been interested in peyote since the 1930s and had become acutely aware of the confusion regarding the plant's botanical classification. His primary interest was the physiological effects of peyote on humans, but he felt his own work and that of others would be more valuable if the botany of peyote and related cacti was clarified. That was my assignment, and it not only became the heart of my graduate work at Claremont but also presented me with research problems for nearly four decades.

Dr. Alles had hoped that the botanical study of peyote could be completed within two years. Immediately upon my release from the army, I went to Mexico to collect plants, but I was unable to take them into California because the state legislature had classified peyote as a narcotic several years earlier, and a state law prohibited possession of peyote even for research purposes. Although Dr. Alles had worked successfully for the passage of a special law by the California state legislature to permit the scientific study of plants classified as narcotics, Governor Knight vetoed the bill. This delayed our peyote research for two years, but in 1959 we at last were given permission to study peyote in California, and my work began in earnest.

The mass of literature on peyote impressed me, but most publications were highly specialized or narrow in their coverage. The most recent attempt at a comprehensive publication that I could find was by Alexandre Rouhier, written in French in 1927. Weston La Barre's *The Peyote Cult*, first published in 1938 and last revised in 1989, and Omer C. Stewart's *Peyote Religion: A History*, published in 1987, are classic anthropological treatments of the peyote religion and will remain important works, but neither scholar attempted an interdisciplinary treatment of the plant like Rouhier's. Despite the considerable amount of information gained about peyote since 1927 and the upsurge of interest in the plant, I could find no other comprehensive overviews. A friend suggested that when I completed my botanical studies of peyote, I should then write an ethnobotanical book about it. The seed was planted, and the first edition of this book was the harvest.

Now, fifteen years later, I have extensively revised the first edition, with the hope that new information and some important stylistic changes will continue to help people understand this remarkable cactus and its significant relationship to Native Americans.

I am much indebted to many colleagues and friends in the United States and in Mexico. I wish to acknowledge especially the inspiration and guidance of the late Lyman Benson, who served as my graduate adviser. I also want to acknowledge the financial assistance provided by the late Gordon A. Alles. The interest of these two men in peyote and their enthusiastic support of my graduate work at Claremont ultimately led to my writing this book.

Other major financial support for my research and travels was provided by the National Science Foundation (GB-2936), the American Philosophical Society (Grant No. 5149, Penrose Fund), and the Cactus and Succulent Society of America. I also wish to thank the University of Arizona Press for bringing about the publication of both editions of this book.

Other persons whom I want especially to thank are Norman H. Boke, Helia Bravo H., Richard C. Brown, Sherwin Carlquist, George Castile, Jane Cole, Christopher Davidson, Anthony Davis, W. A. and Betty Fitz Maurice, Charles Glass, Salvador Johnson, Helen Kinsinger, Jerry McLaughlin, Elizabeth Moore, Shirley Muse, Jerry Patchen, Larry Paynter, David Sands, Stacy Schaefer, Margaret S. Stone, James S. Todd, Carol Wehr, and John Yellowhorse.

The following people, now deceased, also greatly assisted me in my research: Richard O. Albert, Howard Alcock, Edward W. and Dorothy Anderson, Bruce Bowman, Lloyd Brinson, Nancy Frasco, Amy Jean Gilmartin,

Dudley Gold, Hernando Sánchez-Mejorada, Lisa Sergio, David L. Walkington, and Willi Wagner.

Finally, none of my research and writing would have been possible without the support and encouragement of my wife, Adele, to whom I dedicate this book; I thank her for her loving help.

Introduction

The Mexican desert was hot and dry. A few minutes earlier we had seen a small dust devil blow across the gentle plain toward which we were walking. I glanced up at the tall, trim Yucca plants but quickly refocused my attention on penetrating the thorn-shrub thicket whose sharp thorns and spines were a constant threat. We knew that peyote, the divine cactus of Native Americans, should be in this region of San Luis Potosí because the Huichol tribe of the Sierra Madre Occidental have an annual tradition of making collecting pilgrimages to this area they call "Wirikuta."

Sweat trickled onto my glasses as I peered around the clumps of spiny agaves and hechtias, and under the skirts of mesquite and creosote bushes, looking for peyote. Then suddenly I saw one, half-hidden by dry, powdery limestone soil and dead leaves that the wind had blown over it. I knelt down and carefully brushed away the debris, forgetting the heat and my thirst. The plant, about five centimeters across and barely protruding above the ground, was rounded, blue-green in color, and had a flattened top with a depressed center out of which radiated nine furrows. The ribs between these furrows had clusters of long whitish hairs, but no spines. I carefully dug into the soil to remove the plant and found that most of it was hidden in the ground as a massive, turniplike root.

Holding the peyote in my hand, I began to recall the various names that Anglos, Native Americans, and Spanish clergy had given it: "dry whiskey," "divine herb," "devil's root," "medicine of God." Botanists know it as *Lophophora williamsii*. This plant, which has had an association with humans for thousands of years because of the effects of its alkaloids, is revered by some — but cursed by others. What follows is the story of peyote, and the controversial and fascinating relationship humans have had with this unique cactus.

PEYOTE The Divine Cactus

Peyote in flower (*Lophophora williamsii* [Lemaire] J. Coulter)

CHAPTER ONE

Peyote in Mexico

When early humans migrated into the Western Hemisphere, they brought simple tools and elements of a hunting and gathering way of life. Their lives were intimately related to the plants and animals they encountered in their new surroundings, for they provided food, clothing, and protection. In these times, humans also depended on certain plants for medicines and for communication with the spirits, and for thousands of years they used fermented drinks, tobacco, and hallucinogens as an important part of their cultures. These early New World settlers sought plants in their new surroundings that would meet their various needs, and over a period of several thousand years, many such plants were discovered. Modern anthropologists and ethnobotanists have identified one hundred different plant species with psychoactive properties that were used by the early inhabitants of the Western Hemisphere (Furst 1976, 2).

Hallucinogenic plants were employed in divination, curing, meditation, and for the relief of hunger and physical discomfort. According to anthropologist Weston La Barre (1972, 261), some of these mind-altering substances can be traced to the Bronze Age (about 3000 B.C.) and probably to Mesolithic times (about 9000 B.C.), and archaeological evidence in the New World supports this observation. Hallucinogens can be recognized in art pieces from the Colima culture of 2,000 years ago (Furst 1974, 64–66) and in plant remains in northern Mexico extending back to 8500 B.C. (Adovasio and Fry 1976, 95). Apparently the early inhabitants of the Americas brought with them a knowledge of mind-altering plants, and it became an integral part of many Native American cultures.

Over a span of almost ten thousand years, there developed within many groups the ancient shamanistic form of religion in which a specially designated person served as an intermediary between the seen and the unseen worlds as a master of the spirits and as a performer of cures (Furst 1972a, viii). Hallucinogenic plants, including peyote, were in the arsenal of tools by which the shaman bridged the gap between the world of spirits and that of humans. The mystic visionary experience was felt to be direct communication of the shaman or other individuals with the forces or spirits of nature, and in both North and South America, certain peoples used psychoactive plants to facilitate the quest for visions. As some of the nomadic hunters began to settle and practice agriculture, some elements of the paleo-Siberian shamanistic religion were retained. Often they were combined with or modified through interaction with other indigenous cultures, some more dominant than others, and they were later influenced by religious beliefs introduced by the European conquerors (La Barre 1972, 278).

One important shamanistic concept that has been widely retained even into the twentieth century is that of the relationship of sickness to the spirit forces. Many people believe that sickness—and death—are the result of spiritual forces working within the body. A hallucinogenic plant can therefore be a powerful "medicine" because it permits the shaman or individual to remain in tune with the spirit forces and to combat those that are evil. The plants themselves have their own spirits, and the shaman or *curandero* (medicine man) can manipulate them to evoke cures; these plant spirits enable them to transcend the limitations of normal humans and to ascend into the realm of the supernatural, where they can directly battle the sources of witchcraft, disease, and death. Peyote, which has been used in the New World for at least two thousand years, has long been revered for its ability to produce visions as just such a medicine (see chapter 5). Apparently peyote has never been used habitually but only in association with sacred religious ritual. To some people, it is a deity in and of itself (Stenberg 1946, 129).

THE SPANISH CONQUEST AND PEYOTE

The Aztec civilization encountered by Hernán Cortés and his band of adventurers had developed a high degree of sophistication in the use of drug plants to counter diseases caused by evil spirits: various cacti, tobacco (*Nicotiana tabacum*), morning glories (especially *Rivea corymbosa*), mushrooms

(*Psilocybe* spp.), angel's trumpet (*Datura inoxia*), and jimsonweed (*Datura stramonium*). Apparently they also used narcotics and anesthetics in their surgical practices and in the mass human sacrifices on the Great Pyramid (Padula and Friedmann 1987). For example, a fine powder made from the plant called *yoyotli* (*Thevetia thevetioides*) was thrown in the victims' faces to partly anesthetize them just before they were placed on the sacrificial stone (Schendel 1968, 71; Mason and Mason 1987, 61). Some of these Aztec drug plants have been identified and studied, but others have been lost, for we do not know to what substance, plant or otherwise, some of the Aztec words in the few remaining manuscripts refer.

The Aztec civilization, especially its religious use of human sacrifice and plants, horrified the sixteenth-century Roman Catholic mind. The Spanish conquistadors saw the mind-altering drugs of the Aztecs as "pestiferous and wicked" poisons from the devil. After conquering the Aztecs, the Spanish victors, under the direction of church leaders, began a systematic destruction of all traces of the vanquished civilization, including their extensive ethnobotanical knowledge (Huxtable 1992, 7). Juan de Zumarraga, the first archbishop of Mexico, searched throughout the former Aztec empire for information about their civilization and, in an orgy of unparalleled destruction, burned thousands of Aztec documents and other items (Prescott 1855, 1: 101–2). Eventually Spanish authorities in Europe recognized the skills with which the Aztecs had used plants in their medical practices, but by the time they ordered the matter to be studied, in many cases it was too late. Most records of the great Aztec civilization, including information about religious and medicinal plants, went up in smoke. All that remains are a few post-Conquest reconstructions of this New World culture.

European reports of peyote first appeared in the middle of the sixteenth century, but interestingly, the first European book dealing with Aztec plants makes no mention of it. The Badianus manuscript, titled *An Aztec Herbal of 1552*, is a small, hand-written book long preserved in the Vatican Library but recently returned to Mexico. Originally written in the Nahuatl language by an Indian physician named Martín de la Cruz, the manuscript was translated into Latin by his colleague Juan Badiano. Yet even though the book describes more than a hundred medicinal herbs and includes 118 pages of colored paintings, there is no mention of peyote (Furst 1995).

Apparently, the first manuscript reference to the peyote cactus was in the report of the Franciscan friar Bernardino de Sahagún in his *Historia general de las cosas de Nueva España* (General History of the Things of New Spain). Around 1560, Sahagún was working at Tlaltelolco near Teonochtitlán (now Mexico City). With the aid of several Aztec physicians, he attempted to organize Aztec medical knowledge for use in both Old and New Spain. His report, not published until 1829, contains a section on the medicinal herbs used by the Chichimecas, who "discovered and first used the root which they call *péyotl*, and those who ate it and drank it took it in place of wine" (Sahagún 1938, 118). He commented further that "There is another herb, like 'tunas' [prickly-pears or *Opuntia* spp.] of the earth that is called péyotl; it is white and grows in the northern part [of Mexico]. Those who eat or drink it see frightening or laughable visions; this intoxication lasts two or three days and then disappears. It is like a food to the *chichimecas*, which supports them and gives them courage to fight and they have neither fear, nor thirst, nor hunger, and they say that it protects them from every danger" (230).

Peyote was also discussed in the monumental treatment of Mexican herbs by Dr. Francisco Hernández in his *De historia plantarum Novae Hispaniae* (History of the Plants of New Spain). Hernández was sent to Mexico in 1570 by King Philip II of Spain with the task of recovering as much of the Aztec medical lore as possible. Working with educated Aztecs, Hernández spent five years compiling information in three languages—Nahuatl, Spanish, and Latin—on about three thousand plants, most of which had been used medicinally by the Aztecs. He returned to Spain with this mass of information, but it was not until 1649 that a greatly reduced part of it was published dealing with about a thousand plants. He distinguished two kinds of *péyotl* plants, *Péyotl Xochimilcensi* and *Péyotl Zacatecensi*, only the second of which, apparently, was the cactus *Lophophora*. Other translations and versions of the Hernández manuscript included only portions of the 1649 publication; the Lincei edition of 1651 from Rome, for example, included only the first twelve books and did not reach his chapter on peyote.

The Hernández publication referred to "*De Peyotl Zacatecensi, seu radice molli, et lanuginosa*," which may be translated as "Concerning the Peyotl of Zacatecas, or a soft and lanuginous [downy] root," which he described as

> of medium size, sending forth no branches or leaves above ground but with a certain woolliness adhering to it; on which account I could not make a proper drawing. They say it is harmful to both men and women. It appears to be of a sweet taste and moderately hot. Ground up and ap-

plied to painful joints it is said to give relief. Wonderful properties are attributed to this root (if any faith can be given to what is commonly said among them on this point). It causes those eating it to be able to foresee and predict things; for instance, as whether on the following day an enemy will make an attack upon them? or whether they will continue to be in favorable circumstances? who has stolen their household goods or anything else? and other things of like nature which the Chichimecas try to know by means of medication. Those who want to know where the root is hidden in the ground and where it is growing, or whether it will be harmful, they learn by eating another root. It grows in humid places which partake of the nature of lime. (Hernández 1790, 70–71, translated by G. A. Alles)

Probably the most important early medical description of the effects of peyote is that of the physician Juan de Cárdenas, whose work was published in Mexico in 1591 under the title *Problemas y secretos maravillosos de las indias* (Problems and Miraculous Secrets of the Indians). He devoted the last chapter of part one to differentiating between the factual medicinal effects of the "*Peyot*" on the body and mind, and the effects produced by any accompanying "witchcrafts":

In which it is declared whether it can be witch-craft in the herbs and what are witch-crafts.

For us is left to declare that in the Indies it is experienced of the *Peyot*, of the *Poyomate*, of the *Hololisque* and even of the *Piciete*, that if the above mentioned herbs are taken by mouth, many affirm mostly Indians, Negros and dull stupid ignorant people, they imagine and see the devil, who talks to them and tells of things to come. It is now fair to investigate if some herb or root exists in nature the virtue of which could be so effective that by their means we force the devil to come at our call, or by them we divinate of things to come. In reference to this, it is offered to me to answer that there is a part of this in the herb and that there is part that should be attributed to the devil, I declare furthermore, that when one of these herbs or any other that be similar in virtue is taken by mouth or use is made of them the herb itself makes out of its virtue and by its nature three things in the human body, and everything else is illusion and the work of the devil. The effects that the herb makes are the following: In the first place, these herbs are within themselves extremely hot and strong and at the same time are composed of parts extremely subtle, strong, and hot, and that in entering into the stomach the natural heat starts to alter and heat these parts, and in heating them makes the more subtle strong, and hot parts of this herb ascend and distribute in the cerebrum and every part of the body in the form of a vapor. All these extremely strong parts are distributed to the cerebrum and in every way in pores of the body, and they start to heat, perturb and disorganize the animal spirits of the body, taking the man out of his judgements, as is done by wine, the Piciete, and every herb, and even a drink, and any sustaining material strong and vaporous and this is the first effect that the herb or root makes out of its own virtue. The second effect that the said herb makes, is to cause annoying and painful dreams in the man who takes it, and this is caused by those same thick vaporous fumes of the same herb, which although at the beginning may be subtle, on thickening with the coldness and humidity of the cerebrum, so come to cause sleep, not smooth easy and pleasant as the one that comes and is caused out of the soft and humid vapors of sustaining materials, but a sleep horrible and terrifying as is naturally caused by fumes strong and painful. The third effect of the afore-mentioned herbs, or of their painful fumes, is to disturb and disorganize the activities that are in the interior senses of the cerebrum and perturbing those that are represented in the imagination, not with types and forms of things that are enjoyable and entertaining the so mentioned imaginative potency, but instead terrifying and awful things and so they imagine types of figures of monsters, bulls, tigers, lions, and ghosts, that is, painful and horrible things, and nothing less is possible because if the black fumes, strong and heavy are the ones that cause such sleep, and the ones that affect the fantasy, it is understood that they will not represent in the imagination beautiful things pretty, colorful and pleasant, but on the contrary will be wild beasts and horrible things, as the figure of the devil should be represented by the figure of a horrible monster; and so that all of these effects of throwing away the judgement of the one who takes it, to cause horrible sleep, and to represent a species of horrible things in the imagination, it can be made out of its own virtue by every herb of the ones above-mentioned, and therefore, there is reason and cause to be able to make it. The one thing that the herb or root can't make without having a pact and communication with the devil, is what I am going to say now, in the first place the devil to come at the call of the wicked man who searches for him, this is something that the herb cannot do, and it is completely false to say that the herb out of its virtue makes the devil appear. In the second place, to say that, due to the virtue of the herb we know of things to come or secrets that have occurred, it is a notable error that the devil says or declares this, that I understand it, as in effect was told and declared in ancient times by the mouths of the oracles, those false gods, but that the herb out of its virtue could make either one of these things, I have it as a falsehood and lie. (Cárdenas 1945, 243–47, translated by G. A. Alles)

The Cárdenas report is of considerable interest because he discredits the explanation that supernatural forces were the sole cause of visions from these drugs. In fact, he clearly sets forth the concept of the pharmacodynamics of drug absorption in the intestine and its distribution to a site of specific action in the brain. He also correctly observed that there were three biological effects: (1) changes in perception, (2) production of the characteristic visual hallucinatory responses following the early stimulatory phase, and (3) induction of sleep of a non-relaxing nature.

Spanish religious leaders responded quickly and strongly against the use of drug plants by the inhabitants of New Spain. Of course, peyote to Native Americans was neither an instrument of Satan nor an artificial paradise; rather, Peyote was a spirit force or deity that guided them into a real and vital communication with their gods. In contrast, the Spaniards perceived such miraculous powers and communication with God as coming only from observance of the Mass and from miracles performed by the saints. Thus, although the Spaniards believed local inhabitants' accounts of the effectiveness of plants, they concluded that the "miracles" attained through their use could only be the work of the devil. It therefore became a major goal of Spanish religious leaders to stamp out the use of this "satanic gift" (Petrullo 1934, 15).

Roman Catholic clerics established a formidable array of prohibitions against the use of peyote. In 1571 the Inquisition was introduced into Mexico, and by 1620 it was officially declared that since the use of peyote was the work of the devil, all Christians were prohibited from using it (Leonard 1942, 326; also see chapter 9). An early seventeenth-century publication by Fr. Nicolas de León titled *Camino del cielo* (Road to Heaven) presented a series of questions that the confessor should direct to Native American penitents. Some items clearly indicate the widespread fear of witchcraft and other occult activities of the Native Americans:

> Art thou a soothsayer? Dost thou foretell events by reading signs, or by interpreting dreams, or by water, making circles and figures on its surface? Dost thou sweep and ornament with flower garlands the places where idols are preserved? Dost thou know certain words with which to conjure for success in hunting, or to bring rain?
>
> Dost thou suck the blood of others, or dost thou wander about at night, calling upon the demon to help thee? Has thou drunk *peyotl*, or hast thou given it to others to drink, in order to find out secrets, or to discover where stolen or lost articles were? Dost thou know how to speak to vipers in such words that they obey thee? (Brinton 1894, 14)

Another early confessional suggested the following question and the confessor's proper response:

> *Question:* Have you loved God above all things, and adored His Divinity and Majesty above all others, with all your heart, mind, and will? Or have you placed your love in another creature, adoring it, and having it divine, and coming to worship it?
> *Answer:* I have loved Him with all my heart, but sometimes I have believed in dreams, in herbs, in ololiuhqui, and peyote, and other things. (de Alva 1634, 8)

The active prohibition of the use of peyote, and its mention in the confessionals of the Roman Catholic Church, persisted with undiminished fervor into the eighteenth century throughout Mexico and even into what is presently the state of Texas. A small religious manual written in 1760 by Fr. Bartolomé García (1760, 14–15) presents a series of fascinating questions for the penitent:

> Have you killed anyone?
> How many have you murdered?
> Have you eaten the flesh of man?
> Have you eaten peyote?

Thus, García associated use of peyote with murder and cannibalism!

Regardless of the Roman Catholic Church's efforts to prohibit peyote, Native Americans clung tenaciously to the use of the plant because of its great religious and medicinal significance. By the late seventeenth century, some church leaders tried to compromise with indigenous groups where the worship of peyote was still firmly entrenched. In Coahuila, for example, a mission was named El Santo de Jesús Peyotes (Taylor 1949, 81). Elsewhere, Franciscan fathers simply transferred the miracle-working powers of peyote to a calendar saint named Santa Niña de Peyotes. Even today, many Mexican Native Americans practice an unusual mixture of Catholicism and peyotism in which a Catholic priest also serves as a curandero during the night-long ritual that involves eating peyote (Wasson 1965, 164).

Native American Use of Peyote

During the seventeenth and eighteenth centuries, Spanish explorers and missionaries penetrated into the more remote regions of Mexico, often coming in contact with groups of Native Americans who had long been isolated from any other populations of human beings. Peyote and other psychoactive plants were found to be important elements in their religious and social life.

The use of peyote by the Cora tribe of western Mexico (fig. 1.1) was vividly delineated by José Ortega in 1754 in his *Historia del Nayarit, Sonora, Sinaloa, y ambas Californias* (History of Nayarit, Sonora, Sinaloa, and both Californias). His description (1887, 22–23) of their nocturnal dances was translated by Safford as follows:

> Close to the musician was seated the leader of the singing whose business it was to mark the time. Each of these had his assistants to take his place when he should become fatigued. Nearby was placed a tray filled with *peyote* which is a diabolical root (*raiz diabolica*) that is ground up and drunk by them so that they may not become weakened by the exhausting effects of so long a function, which they began by forming as large a circle of men and women as could occupy the space of ground that had been swept off for this purpose. One after the other went dancing in a ring or marking time with their feet, keeping in the middle the musician and the choir-master whom they invited, and singing in the same unmusical tone (*el mismo decompasado tono*) that he set them. They would dance all night, from 5 o'clock in the evening to 7 o'clock in the morning, without stopping nor leaving the circle. When the dance was ended all stood who could hold themselves on their feet; for the majority from the

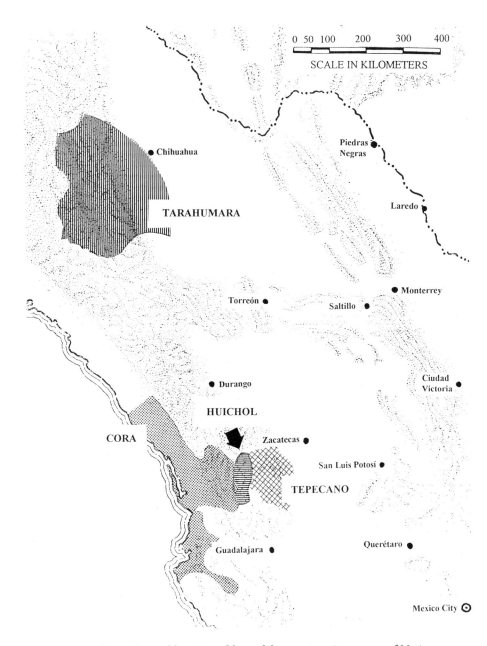

FIGURE 1.1. Map: General location of four of the peyote-using groups of Native Americans in Mexico

peyote and the wine which they drank were unable to utilize their legs to hold themselves upright. (Safford 1915, 295)

The use of peyote apparently was quite widespread during this period, for another early report, this one from the state of Tamaulipas by A. Prieto (1873, 123–24), described the festive occasions during which peyote played a primary role in that part of the north. Fr. Joseph de Arlegui (1851), in his chronicle dealing with the province of Zacatecas, also discussed the ritual and ceremonial use of peyote in which a drink is made "with a root that they call Peyot, which has efficacy not only for intoxicating whoever drinks it, but which makes him almost insensible, benumbing the flesh and deadening all the body" (144).

When the explorer and anthropologist Carl Lumholtz penetrated the wilds of the Sierra Nevada Occidental at the end of the nineteenth century, he found the isolated Coras, Huichols, Tepecanos, Tarahumaras, and other tribes using peyote much as they had for centuries. An examination of the ritualistic use of peyote by some of these peoples illustrates the great religious and social significance that these Native Americans still place on this particular psychoactive plant at the end of the twentieth century.

Huichol Peyotism

The Huichol tribe numbers between ten and twenty thousand people who live on small ranches in the Sierra Madre Occidental on the border of the states of Nayarit and Zacatecas (see fig. 1.1) (S. Schaefer, pers. comm.). This is one of the few Native American tribes whose societal structure, traditions, art, and religion have survived for centuries more or less intact. The Spaniards had little lasting cultural influence in this region, although in the eighteenth century, after several unsuccessful attempts, Spanish troops finally subdued the Huichols to the extent that five missions were established for roughly one hundred years. With the coming of Mexico's independence from Spain, however, the Spanish military and missionaries departed, and the Huichols returned to their former activities (S. Schaefer, pers. comm.).

At the end of the nineteenth century, Carl Lumholtz made two expeditions into the Sierra Madre. He lived with the Huichols for close to ten months and recorded many of their customs and religious practices (Lumholtz 1900; 1902, 2:xiii–xi, 1–290). In recent years, other anthropologists, including Peter T. Furst, Barbara G. Myerhoff, and Stacy B. Schaefer, have studied and recorded the life of these people, who are commonly referred to as "the peyote tribe" of Mexico.

The following account of Huichol religion and ceremonies is based primarily on extensive anthropological studies by Lumholtz (1900, 1902), Furst (1972b, 1973, 1975, 1976, 1978), Muller (1978), Myerhoff (1974, 1978), and Schaefer (pers. comm.). Although Lumholtz's research was separated from that of the others by seventy-five years, the basic patterns and ideas of Huichol peyotism seem to have changed very little.

The Huichol religion is a highly complex one in which many natural phenomena such as fire, air, earth, and water are personified. Fire is the greatest god and existed before the sun; thus, it is called Grandfather and the sun is Father. The goddess who created the world and all that lives is known as Grandmother Growth; she produced all vegetation and is considered the mother of the gods. Her counterpart, Grandfather Fire, is equally antiquated and brought Fire to the people before Father Sun existed. The peyote pilgrimage is led by the burning embers and rekindled fires of Grandfather Fire to the desert and east where Father Sun was born. There are many more gods, most of whom personify the various forces of nature.

Huichols provide their gods with sacrificial animals and all kinds of offerings as prayers for health and good life. Traditionally, the principal animal offered to the gods is the deer, but in recent years cattle and fowl have been sacrificed along with various plant products such as maize (corn), beans, and squashes. To the Huichols, all staple foods are gifts from the gods, so they will not partake of any new crop until there has been an appropriate ceremony or feast in which part of the food has been offered to the gods. Maize is especially revered, and, interestingly, the Huichols regard it and peyote (called *híkuli* or *híkuri*) as derived from deer.

Not only does all food come from the gods, but the Huichols also believe that diseases are caused by gods who come at night to make people ill. Through dreams and other mystic practices, the Huichol shamans learn the nature of these diseases and combat them, working to help the patient regain equilibrium and health. They often use plants as medicinal remedies, and peyote is an important plant in their pharmacopeia. It is widely respected for its healing properties, and is even used to relieve pain caused by scorpion stings.

The Huichols engage in a variety of ritualistic rites, such as for making rain and celebrating successful crops. Peyote is ever-present in Huichol beliefs and traditions, and it plays an important role in many ceremonies, especially those which revolve around the peyote pilgrimage and the planting and harvesting of crops. Although the peyote pilgrimage and associated ceremonies are traditionally performed during the dry season, from around November to May, the pilgrimage and peyote itself are interrelated with nearly all other ritual events, including rainy-season ceremonies. The relationship of this hallucinogenic cactus to the Huichol religion is complicated and difficult to comprehend, but basically the main desire of the Huichols is to have rain in order to raise maize, their primary food. In addition, "peyote is meaningful because it unifies family and community members. . . . [I]t serves an enculturative purpose, instilling and reinforcing the importance of cultural beliefs and values, as well as a collectively shared worldview" (Schaefer 1996, n.p.).

Huichol tradition holds that maize was once the deer, which had been the chief source of food in earlier times. Thus the deer represents sustenance, and legend says that both peyote and water sprang from the forehead of a deer. Peyote, in fact, is believed to be the original ear of maize, and the deer antler is the original peyote. This confusion makes sense only if one realizes that the Huichols believe that maize, deer, and peyote are one and the same thing: "corn [maize] is deer (food substance), and híkuli [peyote] is deer (food substance), and corn is híkuli" (Lumholtz 1900, 2). As one Huichol man said, "Peyote is the crossing of the souls, it is everything that is. Without peyote nothing would exist" (quoted in Schaefer 1996, n.p.). Muller (1978, 84) suggests that this "sacred peyote-deer-corn trinity" represents the main stages of unrecorded Huichol history, with the peyote and deer symbolizing their hunting-gathering period, and maize the more recent agricultural stage. The Huichols believe that the deer-god left the peyote plants in his tracks when he first appeared in the land where peyote grows. Thus, peyote is the plant of life—the life of the deer and of maize. Furthermore, peyote is also the drinking bowl of the greatest of all gods: Grandfather Fire. The Huichols must therefore obtain peyote and bring it to Grandfather Fire every year or it will not rain, nor will they have maize or deer.

The Huichols also believe that "peyote is for learning," that "those with strong hearts will receive messages from the gods" (quoted in Schaefer 1996, n.p.), and in October they engage in a sacred pilgrimage to Wirikuta, the land where peyote grows. A successful trip will ensure the growth of their maize, their children, and all of nature. As many as twenty or more persons participate in the three-hundred-mile trip to an area near Real de Catorce in the state of San Luis Potosí, and before the use of cars and buses, the pilgrimage lasted about forty days. The ritual demands that participants not wash or eat salt and that they endure long fasts. They are pleased to go, however, because their involvement in a successful journey will ensure adequate rains, good crops, health, and happiness for themselves and the others within the tribe.

One of the major preparations for the pilgrimage is the ritual of confession and purification. This remarkable ceremony requires pilgrims as well as their family members staying home to declare publicly their sexual activities. Novice pilgrims must confess all their sexual transgressions from the beginning of adulthood to the present; on subsequent pilgrimages, they confess only extramarital affairs that have occurred since the first pilgrimage in which they participated. During the ceremony, no one is to show any jealousy or anger (Furst 1976, 116), and as each person gives his or her account, the leader ties a knot for each incident in a confession cord made of *ixtle* (*Agave lechuguilla*). After all have confessed, he throws the cord into the fire, which symbolizes that all are purified (Myerhoff 1974, 131–36; Furst 1976, 116).

The night before leaving for the land of peyote, the pilgrims take a bath, offer prayers, and sleep with their families in the temple. Sexual abstinence is required once the journey begins, even if one's spouse is on the pilgrimage, too.

The following morning they offer additional prayers and sacrifice five tortillas to the fire. All the pilgrims, including the shaman or leader, carry sacred gourds containing small packages of tobacco to be used ceremonially when the group arrives in peyote country. This tobacco, called *yé*, is *Nicotiana rustica*, a species with exceptionally high nicotine levels (Siegel, Collings, and Díaz 1977, 16–22). Taken in conjunction with peyote, the tobacco "may in fact facilitate (a better term might be 'enhance') the transition from the physical to the metaphysical worlds traveled by the pilgrims" (Schaefer 1996, n.p.).

Before leaving the village, new pilgrims are blindfolded, and then the group departs carrying lighted candles. Whether the journey is on foot as in the early days or by modern car or bus, the route is carefully prescribed by the leader because the group must stop and perform ritual activities at certain critical places. One of the most important of these sites is Tatéi Matiniéri, the sacred water holes which were once part of a lake that dried up long ago. Tatéi Matiniéri is where "Our Mother dwells," and the Huichols leave offerings to her in and around the water holes (Furst 1972b, 166–70). The pilgrims also fill all their water containers here, to be used in the peyote area and later back home on their crops and when eating peyote. Everyone is ritually washed with the water, which is poured over their heads, and then the blindfolds of the new members are removed as they look ahead to the land of peyote.

The peyote seekers proceed to Wirikuta, timing their arrival for early morning, when all participate in building a fire at a place designated by the leader. Each pilgrim "feeds" the fire dry sticks and creosote (*Larrea tridentata*) branches, for Grandfather Fire will protect them while they are in Wirikuta, also known as the place of the five-pointed deer. This region is the most sacred to the Huichols, not only as the place of peyote but also as the place they come upon death, before finally settling in the west.

The Huichol myth that describes the discovery of peyote also explains another part of the ritual at Wirikuta:

> Long ago, when the forefathers of the Huichols first arrived in the country where the hi'kuli [peyote] now grows, they saw a deer, and allowed him to go five steps, when he disappeared. When they came closer to the tracks, they discovered that each footprint was a hi'kuli. All together, there were five—one for each footprint.
>
> They shot arrows at every hi'kuli without hurting it, two arrows above each, and in such a manner that the end of one arrow pointed to the east, and the end of the other to the west. At the place where the deer disappeared a large hi'kuli was found, which was called Pa'li or Wapa'li. After a while they proceeded to take up their arrows and put them into their quivers. Only the two arrows which they had shot above the big hi'kuli remained, because Great-grandfather Deer-Tail ordered them to leave them. Then they sat down and ate hi'kuli. Tama'ts Palisi'ke remained on the high mesa where hi'kuli first appeared, and there he may be seen to-day in the form of an altar. He is the principal altar,—a big hi'kuli. (Lumholtz 1900, 18–19)

As the sacred fire burns, the peyote seekers arrange themselves in a circle around it and pray for a successful "hunt" of the Deer-Peyote. About midmorning, the leader signals for the hunt to begin. Each pilgrim tests his or her bow and draws an arrow while making a final circuit around the fire, at the same time adding a little more "food" to the fire and praying for protection. The pilgrims then fan out and creep toward the distant hills, crouching low and scanning the ground. Suddenly the leader freezes in his tracks and motions for the others to join him. Pointing to a small cluster of peyote beneath a bush, he takes careful aim and shoots an arrow at it. It strikes the ground just to the left of the clump; another arrow is fired, striking to the right, and then quickly a third arrow is shot just in front of the plant. The leader now rushes to the peyote and places a fourth and final ceremonial arrow so that the Deer-Peyote is surrounded on four sides by arrows in each of the world quarters.

If the peyote top has only five ribs, the pilgrims are especially pleased, for five is the sacred number of completion corresponding to the five-pointed deer and the five sacred colors of maize. As the peyote seekers gather around the "dying deer," they pray for it not to be angry, as they have brought offerings of tobacco, water, and tortillas. The leader carefully digs the dirt away from the clump and cuts it off at the base so that part of the root remains to grow again. Each plant is then cut into five parts, and others are harvested in the same manner so that all the pilgrims can eat a piece as they squat or kneel in front of the leader.

Near the spot where each peyote plant is found, the pilgrims place their tobacco-filled gourds (Furst 1976, 122–26). Other offerings are piled in front of the holes and set afire, including miniatures of the famous Huichol *nearikas*, or yarn paintings. These paintings, created in wool and wax on boards, are inspired by the visions experienced when eating peyote, and by the pilgrimage itself. As the flames consume the offerings, the pilgrims pass fresh portions of peyote which they touch to their foreheads, eyes, throats, and hearts before placing it in their mouths and eating it. Often the peyote seekers dance and sing as they eat the peyote. Later the leader instructs them to search for more peyote, and the pilgrims walk slowly through the brush looking for the plants, which they carefully dig and cut, speaking quietly and tenderly to them as they place them in their baskets or bags.

Late in the afternoon the peyote seekers return single file to camp, where they walk ceremonially around the fire, weeping and offering thanks for the protection it has provided them. That night they eat a large number of peyote buttons while they sing, dance, and tell ancient stories. All are happy if the hunt has been successful.

Furst (1971, 183) noted that other cacti may confuse pilgrims in their search for peyote at Wirikuta, especially those with an "impure heart," meaning they had not been properly purified prior to the journey. The Huichols believe that eating this cactus is "very dangerous." Furst has identified the "false peyote," called *tsuwíri*, as *Ariocarpus retusus*, a fairly common spineless cactus that grows throughout northern Mexico.

The pilgrims may collect plants for up to three days, celebrating each night, but soon it comes time to leave the land of Wirikuta. The fire is extinguished with some of the sacred water, and the pilgrims shout their farewells to Peyote. Before leaving, they paint their faces yellow, using a sacred paint obtained from the roots of a shrub called 'uxa (*Mahonia trifoliata*) (Bauml, Voss, and Collings 1990, 99–101). They then depart on foot, single file, blowing their ritual horns. Although sad that they must leave, they are pleased that their baskets are full of peyote.

As the peyote seekers near home, they begin a deer hunt that lasts three to six days. They must kill enough deer to ensure that it will rain the next season and to free them from the requirements of continence, fasting, and using no salt. During the hunt they subsist solely on peyote and sleep very little. The deer meat is cut up, cooked, and threaded onto strings where it is allowed to dry and harden. Some of the peyote is also strung on cords and placed with the deer meat in the temple to be used the following year. When the deer hunt is successfully completed, the pilgrims are allowed to bathe for the first time since leaving on the journey to Wirikuta.

The Fiesta of the Peyote, or Híkuli Feast, is the most important of all Huichol ceremonies and is held in January for a period of three days (Lumholtz 1902, 2:268–80). Prior to the festival, men and women clear the temple field or patio, prepare elaborate costumes, and make *tesvino*, their beer made from maize. After grinding sprouted maize kernels, they add them to large containers filled with water to slowly simmer on an open fire overnight. The feast begins at dusk the following day, when three fires are built around the patio: a temple fire (which has a special "pillow" of green wood for Grandfather Fire), one to the east to guard the dancers, and a third to the north for "visitors" from the underworld, thus providing both light and protection.

The peyote seekers who participated the previous year in the pilgrimage carry tamales in pouches slung over their shoulders, circling the temple fire and depositing the tamales on a blanket by the altar. Some of the food is offered first to the fire; the rest is then distributed among those present. At the same time, everyone drinks some of the water brought from the sacred water holes. The leader and pilgrims enter the temple and pray, surrounded by symbolic items such as stuffed animals, crucifixes, jars of tesvino and sacred water, and feathers. One symbol of particular importance is a stuffed squirrel, elaborately decorated with feathers and beetle wings. This animal is believed to see more clearly than people; it guards them against evil spirits and guides the peyote seekers on their journey. Prayers center around the journey the previous October and usually last until midnight, when the pilgrims leave the temple and begin to sing and dance. The leader sits in front of the temple fire, facing east with assistants on either side. Both sexes dance while the leader or one of his assistants sings. No drum is used, and the dance, which is not continuous, is described by Lumholtz (1902) as "a quick jumping walk with frequent jerky turns of the body" (2:274). Each dancer carries a bamboo stick which represents a serpent, and as they dance, the stick is rapidly thrust in all directions.

At noon on the second day, the participants stop their dancing and singing long enough to paint their faces with symbolic designs in yellow, which most often represent peyote, flowers, clouds, and maize. The dancing and singing, as well as the drinking of the sacred water and the eating of peyote, continue in a more or less uninterrupted fashion until sunrise on the third and last day. The rising of the sun denotes the beginning of a day of great rejoicing; the participants begin to drink large amounts of tesvino and a native brandy, and soon many are drunk. This final day reaches its culmination in *Rarikira*, the maize-roasting ceremony. A woman is selected to toast the maize that is brought to her by the peyote pilgrims, who ceremonially circle the fire and then sit down to remove the kernels from the cobs. Each pilgrim first offers five kernels to the fire and then hands the remaining kernels to the woman to toast and make into *esquite*, the Huichols' basic maize dish. The esquite, deer meat, and broth are shared by all the festival participants, and the ceremony ends with the eating of this ritual meal.

The Huichols and others orally ingest peyote, chewing the fresh or dried tops of the plants and then swallowing them. There have also been reports of the ritual use of peyote enemas (Furst and Coe 1977; de Smet 1983; de Smet and Lipp 1987). In fact, Hernando Ruíz de Alarcón wrote of such activities in 1629: "Others for the said illness of fevers use enemas [*aiudas*] using as herbs [*simples*] at times the said *ololiuhqui* [*Turbina corymbosa*] or *peyote* and at times *atlinan* or other herbs; and whether it be one or the other, the method is to grind it up and dilute it in cold water and to put it in as an enema" (1898, 218; my translation).

In view of the documented use of peyote and other psychotropic enemas in central Mexico, this practice may have reached the western Sierra Madre with the Nahuatl-speaking Native Americans transplanted by Spaniards to the western frontier beginning in the sixteenth century. Furst and Coe (1977, 91) describe a report by Tim Knab, who had been shown a Huichol "peyote enema apparatus" made of the bladder and hollow femur of a small deer. Knab also reported that peyote was prepared by the shaman and injected rectally into an elderly woman. However, de Smet (1983), commenting on Knab's report, stated that its "actual use has not been witnessed" (145). Such use certainly is not found in any other reports of ceremonial use of peyote by Native Americans, although Furst (1976, 28) suggests that those who take peyote rectally may do so because of weak stomachs that cannot tolerate peyote's bitter, nauseating taste. Certainly many other hallucinogens are taken rectally (de Smet 1983), so one cannot readily dismiss Knab's report. In fact, La Barre (1979) believes that Chavin pottery from Peru suggests that infusions of the hallucinogenic San Pedro cactus (*Echinopsis pachanoi*) were taken rectally by Native Americans more than 2,500 years ago (34).

Tarahumara Peyotism

The Tarahumaras of Chihuahua are an important remnant of the semi-agricultural people who have inhabited the northern portion of the Sierra Madre Occidental for hundreds of years with little apparent cultural change. About 50,000 of them live scattered throughout the hill and plains regions west and south of the city of Chihuahua (see fig. 1.1). Although the first Spanish contact with the Tarahumaras was in 1607, and relations with the Mexican government have been maintained to at least some degree ever since, certain aspects of Tarahumara life have not changed. As with the Huichols to the south, peyote has long been important to the Tarahumaras. At the end of the sixteenth century, Francisco Javier Alegre reported in his *Historia de la provincia de la compañía de Jesús de Nueva España* (History of the Province of the Company of Jesus of New Spain) that the Tarahumaras used powdered peyote for treating wounds (1958, 465). Later observers found that peyote, along with several other cacti, were highly revered and objects of elaborate ceremonies (Bye 1979a, 27–41).

Carl Lumholtz (1902, 1:326) reported that the greatest Tarahumara healers were those who made a specialty of the *híkuli*, or peyote cult, for few plants other than peyote were employed for medicinal purposes. Bye (1979a; 1985), on the other hand, has compiled an extensive list of Tarahumara medicinal plants, including a number of cacti.

Wendell Clark Bennett and Robert M. Zingg (1935, 136) in their important work on the Tarahumaras, stated that not only do the Tarahumaras drink powdered peyote in water to give health and long life and to purify body and soul, but they also chew and then apply it externally to treat snakebite, bruises and wounds, burns, fractures, constipation, and rheumatism. Bennett and Zingg further suggested that the Tarahumaras' use of peyote "resembles an elaborate curing ceremony rather than a cult. There is nothing to suggest a society centered around peyote-eating. . . . The group of peyote-eaters does not involve any exclusiveness, requirements, or ritual pertaining to individuals. The peyote ceremonies are not given for the pleasure of eating the plant, but to cure some disease" (294).

In a more recent study of Tarahumara medicinal practices, Bye (1985, 77–83) noted that preventive health care is usually ceremonial in nature, with a variety of plants, including cacti, employed "to foil sorcerers and evil beings." The hallucinogenic plants, including peyote, require special treatment because they are so powerful in effecting cures; they are even said to sing to the Tarahumaras (Bye 1979a, 26).

Bennett and Zingg (1935, 366–67) believed that although the peyote, or *híkuli* (also spelled *híkuri*), cult was once very important to the Tarahumaras, it had become much less important and not so intricate as among the Huichols. It does seem clear, however, that the two groups have shared many of the concepts of peyotism; peyote has the same name in both languages, and parts of the collection pilgrimage and fiesta are similar.

The Tarahumara pilgrimage, in contrast to that of the Huichols, often combines trading and other commercial aspects, and there are no restrictions on what foods may be eaten during the pilgrimage or what time of year it is made. However, on their journey, and when the pilgrims reach peyote country, they typically eat only their usual traveling food, *pinole* (powdered maize added to some type of liquid) (Bennett and Zingg 1935, 292). Apparently their main collecting area is in northeastern Chihuahua along the Río Conchos, which is not too far from Presidio, Texas, and the Rio Grande. Peyote is known to occur naturally in this area north of the border, but there are no known locations for it in this relatively unexplored region of Mexico.

At the collection area, the Tarahumaras erect a cross around which they place the first peyote collected so that these plants can tell the pilgrims where others are. They eat the next plants that are collected and then quietly lie down and go to sleep. The following day they return to the field to collect more peyote, singing joyfully while doing so. That evening a fire is built near the cross, and the pilgrims dance. The harvest lasts for several days, with some pilgrims digging plants while others sleep or dance. The Tarahumaras claim that the peyote sings beautiful songs while they are in the field to help the pilgrims find them.

As soon as the peyote pilgrims return home, there is a festival to celebrate their success. The peyote is placed upon a blanket under a cross, and the blood of a freshly slaughtered sheep or goat is sprinkled on it. The meat of the sacrificial animal is then eaten, and all dance around a large fire throughout the night. They also consume large quantities of tesvino and fresh peyote tea. There is much ritual involved with the festival, and the dancing is accompanied by singing and "rasping" by the leader, or shaman. The rasping is accomplished by rubbing a smooth "rasping stick" against a notched stick placed on half a gourd inverted over a peyote plant; the gourd thus serves as a resonator while the leader evenly strokes the sticks together. Late in the ceremony the leader often performs cures with his rasp and "doctors" any who may need treatment. Then all participants wash carefully and the festivities are over.

The Tarahumaras are also famous in Mexico as great distance runners, whose races have religious significance. Contestants run barefoot and naked except for loin cloths and little leather pouches containing peyote at their sides. Each athlete eats peyote as he runs so that he will feel less pain and have greater endurance (Roseman 1963, 50–52).

Other groups of Native Americans in Mexico also have a long history and tradition of peyote usage. The Cora of western Mexico have a unique passion play that incorporates both Native American and Roman Catholic ideas. The "Borrados," or Judeans, who are the forces of evil, give themselves strength in an all-night dance by drinking a mixture of cornmeal mush and bits of peyote (Aldana E. 1971).

Mexican peyotism found its way north across the Rio Grande and into what is now the United States prior to the nineteenth century, according to J. S. Slotkin (1955), who published early records of peyote use in the Southwest, the Gulf of Mexico area, and the southern plains states. Additional information was published later by Stewart (1974; 1987, 45–53) and Morgan (1983b, 73). Apparently this old peyote complex never became firmly established in the United States, although some elements of it, including the eating of peyote, were incorporated into modern Plains peyotism. Other U.S.

tribal rites, such as the Ghost Dance, also borrowed from Native American ceremonies in Mexico.

Historically, the peyote cactus has been used for millennia by Native Americans within Mexico. The Mexican form of peyotism, first encountered by the Spanish conquerors, persists among several isolated tribes in northern Mexico and has been a widespread and elaborate community cult, in contrast to the more personalized form of peyotism that developed in the United States during the late nineteenth century (see chapter 2). This old peyote complex of Mexico emphasizes the role of the shaman, who through his or her use of peyote and other powerful plants has been relied on to cure those who are ill or who are encountering evil spirits. The Spaniards learned that they could not successfully prohibit the use of the plant, and accommodation or compromise was the result.

Although there is archaeological evidence that, as long ago as 5000 B.C., Desert Culture hunter-gatherers in the Trans-Pecos region of Texas knew of peyote (and also the highly toxic, but psychoactive, seeds of *Sophora secundiflora*, a small flowering tree in the family Fabaceae) (Furst 1989, 386–87), it is from the peoples of northern Mexico that Native Americans of the United States learned the way of peyote. This knowledge produced a new pan-indigenous religious cult that ultimately extended north into Canada. The following two chapters will deal with the origin and diffusion of modern U.S. peyotism, as well as the peyote ceremony itself.

Chapter Two

Peyote in the United States

As described in the previous chapter, the peyote cactus has been used for several centuries by Native Americans in Mexico and, to a limited extent, by some indigenous groups north of the Rio Grande. In the second half of the nineteenth century, a new form of peyotism originated in the southern United States and rapidly spread throughout the newly created Indian reservations. This new type of peyotism, which could appropriately be called the modern peyote religion, differs from the old peyote complex of Mexico and has appealed to large numbers of Native Americans in the United States because its ritual contains elements of traditional culture as well as certain aspects of Christian symbolism. How did such a different form of peyotism arise, and what conditions led to its eager acceptance by Native Americans? Answers to these questions are unclear in some respects, but Native American and European American accounts of the origin of this new form of peyotism, and a summary of Native American conditions in the nineteenth century, provide considerable illumination.

Native American Traditions

Most tribes that include members of the peyote religion have at least one traditional account of its origin. Some stories begin with the account of a battle and derive certain peyote paraphernalia from weapons of war, whereas other accounts relate an experience in which a member of a hunting party becomes lost. Usually a spirit-force named Peyote comes to a young man or woman in dire distress; in some stories the individual even dies, is transported to a spiritual realm where the peyote message is taught, and is then returned to earth so that he or she can teach all the Native Americans the new religion. Most accounts emphasize not only the individual's nearness to death, but his or her miraculous recovery through the eating of peyote. Thus, from the soil comes a spirit-force in the form of a plant that quickly heals—and teaches. Other aspects of the accounts—loneliness, hunger, thirst, despair—can be seen as symbolic of the fate that Native Americans feel have befallen them at the hands of later settlers from Europe. With the intervention of peyote, however, there is hope and the promise of a better future.

A typical and widespread version of the origin of peyotism is found in a narrative by the Delaware prophet Elk Hair, in which a young woman becomes lost and encounters peyote.

> A long time ago a group of Indians went hunting, taking with them a young boy. Wanting to prove that he too was a man, the boy left his companions to hunt alone. After a successful day the hunting party went back to the camping ground, but the boy failed to return. At that time there were many wild beasts in our land, and the people worried about the child. They searched for him many days, but did not find him. At last his sister, his only relative, decided to search for the boy herself. She decided to look for him in the west.
>
> She wandered about for many days without discovering any trace of her brother, and finally all hope left her. Grieving, she said to herself:
>
> "He is gone. Now I don't care where I go. I don't care what becomes of me."
>
> One morning she began the search early. She was weak from lack of food and water. Coming to a lake, she lay down and prayed to God, saying, "I don't care what becomes of me now, but I hope that God will let me see my brother once more before I die."
>
> She stretched herself out to die, her head to the east, her feet pointing to the west, her stretched-out arms pointing to the south and north, saying to herself, "I don't care to live anymore. Since I can't find my brother, I will die."
>
> All at once she began to reach down several inches in the mud and water with her fingers and felt something cool. At the same time she saw a man standing before her who said to her:
>
> "Here, what is the use of worrying? Look at me! Your people are safe, I

am taking care of them. Your brother is safe. He is still living. If you want to see your brother, look to the west."

She did so, and saw her brother a very long distance away. At the same time something cool touched her hand again. The man disappeared. She looked at what she had in her hand which spoke to her, saying:

"It is I, Peyote. Now you can drink this water. You have had nothing to drink or eat for a long time. Drink this water and you will feel well. Now, eat what you have left in your left hand. Sit down and think about yourself. Think about being happy in this world. Don't worry about your brother. He is safe."

She ate what was in her left hand and she saw her brother again. He said to her, "I am safe. Don't worry about me." He then disappeared. Then Peyote spoke to her again, instructing her how the plant in her left hand was to be used.

"When you get back to your people, show them what you have in your left hand. It is my power put here by God. Use it the way I teach you. Use it to keep well and to keep from worrying in this world. Either drink it or eat it."

The girl went back to the village and told her people.

That is how the Indians discovered Peyote. (Quoted in Petrullo 1934, 38–40)

Interesting variations of the hunting account occur in a version by Albert Hensley, a highly educated Winnebago of Wisconsin, who tells the legend this way:

I know the story about the origin of the peyote. It is as follows: Once in the south, an Indian belonging to the tribe called Mescallero Apache was roaming in the country called Mexico, and went hunting in the high hills and got lost. For three days he went without water and without food. He was about to die of thirst but he continued until he reached the foot of a certain hill, on top of which he could find shade under a tree that was growing there. There he desired to die. It was with the greatest difficulty that he reached the place and when he got there, he fell over on his back and lay thus, with his body stretched toward the south, his head pillowed against something. He extended his right arm to the west and his left arm to the east, and as he did this, he felt something cool touch his hands. "What is it?" he thought to himself. So he took the one that was close to his right hand and brought it to his mouth and ate it. There was water in it, although it also contained food. Then he took the one close to his left hand and brought it to his mouth and ate it. Then as he lay on the ground a holy spirit entered him and taking the spirit of the Indian carried it away to the regions above. There he saw a man who spoke to him. "I have caused you to go through all this suffering, for had I not done it, you would never have heard of the proper [religion]. It was for that reason that I placed holiness in what you have eaten. My Father gave it to me and I was permitted to place it on the earth. I was also permitted to take it back again and give it to some other Indians.

"At present this religion exists in the south but now I wish to have it extended to the north. You Indians are now fighting one another, and it is for the purpose of stopping this, that you might shake hands and partake of food together, that I am giving you his peyote. Now you should love one another. Earthmaker is my father. Long ago I sent this gospel across the ocean but you did not know of it. Now I am going to teach you to understand it." Then he led him into a lodge where they were eating peyote. There he taught him the songs and all that belonged to this ceremony. Then he said to him. "Now go to your people and teach them all that I have told you. Go to your people in the north and teach them. I have placed my holiness in this that you eat. What my father gave me, that I have placed therein." (Quoted in Radin 1923, 389–99)

Hensley identifies the recipient of the message as a young Mescalero Apache man who had his spirit taken away by Peyote, the son of Earthmaker (the Great Spirit). The spirit-force Peyote comments that this religion is already in the south and that Native Americans should love one another. Both historical fact and part of a new ethic are thus presented.

A final example, and in some ways the most interesting story, is told by Howard Rain of the Menomini tribe from Michigan. Even though the basic hunting story still exists, it has a strongly Christian theme.

> There was a family, they had ten daughters and one boy. And this old man, he knows that he is going to die. This old man was a chief of a whole tribe, and he have his son to be a chief. He said, "I'm going to go,

and you take my place. Take care of this [tribe]." And the boy began to think; he wasn't good enough to be a chief and look after his tribe. So he didn't know what to do; he's poor. So he went out hunting, one time. He went out hunting; he goes to that place for a while; he ain't supposed to get lost, but he got lost. He was lost for about four days. He began to get dry and hungry, tired out; so he gave up. There was a nice place there—there was a tree there; nice shade, nice grass—and he looked at that place there; it would be a nice place for him to die. So he went, lay himself down on his back; he stretched out his arms like this [extending his arms horizontally], and lay like that. Pretty soon he felt something kind of damp [in] each hand. So he took them [damp things], looked at them. He got up; that stuff was kind of fresh, you know. "Well, as long as it's kind of juicy, I might just as well take it. I'm going to die anyway; maybe I could live a little longer," he thought. So he ate it—he had two of them in each hand. So he took them, and after he took them, then he passed away. Just as soon as he—I suppose his soul—came to, he see somebody coming on clouds. There's a cloud; something coming. That's a man coming this way, with a buckskin suit on; he got long hair. He come right straight for him, it's Jesus himself. So he told this boy, "Well, one time you was crying, and your prayers were answered that time. So I come here. I'm not supposed to come; I said I wasn't going to come before two thousand years," he said. "But I come for you, to come tell you why that's you [are] lost. You ain't lost. Oh, it's just a little ways, here. But we're going to bring you something, so you can take care of your people. That's what you're crying for; you don't know how—how you're going to take care of your people. So we're going to give you that power to do it. But we go up here first." So they went up a hill there. There's a tipi there, all ready. So Christ, before he went in it offered a prayer. So they went in there. Then he showed him the [ritual] ways; the medicine, how to use it, he gave him the songs them songs we're using—but that's why, see [that] we don't understand them words [of the songs], you know. So he teach him to go through the meeting, just the way I had it over there [when he led the tipi meeting]. So now, after they got through. "Now you can go back. Take this medicine along, over there. Keep your peoples. Whoever takes this medicine, he will do it in my name." So that's how it represents almost the first beginning. (Quoted in Slotkin 1952, 573)

Thus, Christ Himself came to give the peyote ritual to the Menomini, a tribe that incorporates a considerable amount of the Christian symbolism into their ceremonies.

A Comanche account of the spread of peyotism was published by David P. McAllester (1949). The story tells about a battle between the Comanches and Apaches in which all of the Comanches were killed, including a young man who had fought particularly well. The Apaches took the dead warrior's belongings and went to have a peyote meeting. In the middle of the meeting the young scalped Comanche entered the tipi. The shocked Apaches sat in stunned silence for a moment, and then the dead Comanche spoke:

> "You people do not understand peyote power. I, a Yú· 'Ta?, know its power."
>
> He said to them, "You have no power, you do not understand it. After you wore me down completely you stung me to death and took everything I had and I just came to tell you that is how it happened. Now I will tell you something more. Look here, now you people. It is midnight." And he told them then:
>
> "You people, smell the smoke—that is the smoke of Comanches coming to see you. They are coming to you. Smell their smoke. There are seven Comanche. It will be four days before they arrive here." And he instructed them:
>
> "When those seven arrive you are to have a peyote meeting. When you have the meeting I command you to give them my bow and this peyote. They shall take these things from you with them."
>
> One of the Karisu [Carrizo] Apache spoke up in good Comanche and answered him. "Yes, we will do for your people what you request."
>
> Then the young man told them he was ready to go. "After I have taken seven steps from the door I will stop and yell several times. When I have taken the seventh step I will yell four times. On my fourth yell I want you to sing." (Quoted in McAllester 1949, 15)

The Apaches sang a peyote song and the young man disappeared. Four days later the seven Comanches arrived as the dead brave had prophesied, and together they held a ceremony using the dead warrior's weapons. A Comanche girl captive interpreted, and from them the Apaches learned the proper peyote ceremony. This account is particularly interesting because although the Apaches already used peyote, they learned the "correct" ritual from the Comanches through a dead brave and a captive girl.

These legendary accounts provide Native Americans with a historical and cultural explanation of how they came to receive the great "medicine" or spirit-force Peyote and how it spread among the tribes.

POSSIBLE ORIGINS OF PEYOTISM

The actual origin of modern peyotism within the United States is still somewhat unclear, but Stewart (1984, 270–71) reports that peyote—and Christianity—were known to both the Lipan and Mescalero Apaches by 1770. The Lipan Apaches later interacted with the Kiowas, Comanches, and other Native Americans living in Oklahoma; from this western Oklahoma center the modern peyote religion evolved and spread in all directions (Stewart 1987, 51).

It is important to remember that the old peyote complex of Mexico (see chapter 1) was a cult involving community-wide dancing ceremonies in which the plant was used to assure tribal welfare, to give protection, to produce visions for supernatural revelation, to alleviate hunger and fatigue, and to heal sickness.

Apparently there were three major influences that may have produced modern peyotism: (1) the Kiowa ceremony of the sacred stones, (2) the old peyote complex of Mexico, and (3) mescalism. None of these practices are mutually exclusive, and any of the three—or combinations of them—could be valid explanations of the origin of peyote ceremonies found in the United States.

The Kiowa Ceremony of the Sacred Stones

Anthropologist Ruth Underhill (1952) proposed that the modern peyote ceremony is based on an old Kiowa ceremony in which worshippers sat within a tipi in a circle around a crescent-shaped earthen altar that originally held sacred stones with spirit-power. Underhill suggested that the stones were later replaced by a large peyote and that other common ritual features of this Plains ceremony, such as the use of white sage and cedar, the eagle-feather fan, the drum, and the rattle, were also retained in the new peyote ceremony.

The Old Peyote Complex of Mexico

Only the old peyote complex is found in Mexico, and it seems unlikely that modern peyotism was actually invented at the border by Lipan Apache Indians. According to J. S. Slotkin (1956, 34–35), the ritual use of peyote reached the Great Plains area of the United States from various sources, including the gulf and southwestern tribes. This theory of Slotkin and other data suggest a developmental sequence from (a) the old peyote complex that used peyote in community festivals which involved elaborate religious mythology and ritual, to (b) the Lipan Apache rites of the eighteenth century in which the tribal-dancing rite was changed to a quiet ceremony including prayer, singing, and contemplation, to (c) a further modification of the quiet ceremony or "religion-like rite" as it spread throughout the southern Plains. This probable sequence of events is well documented by Stewart (1974, 214–15; 1987, 45–67), who emphasizes that for nearly a century, Apaches were in frequent contact with both Christian missions and Mexican peyote users. To these people, Christianity and the peyote rites were equally strange, so it is easy to see how they might have created a single ritual complex combining practices and symbolism of both.

The modern peyote ritual incorporates several ceremonial procedures of both the Tamaulipan and Tarahumara tribes of Mexico, such as an all-night ceremony around a fire, the eating of food along with peyote, the use of a drum, the significance of the number four, the use of tobacco in cigarettes as a sign of friendship, and an emphasis on the moon and the east. This borrowing and sharing of elements from Mexico probably began with the Lipan Apaches living in missions about 1770 (Stewart 1944, 64; 1987, 40–41). It is remarkable that the modern peyote ritual is so similar among the many tribes in the United States, with all groups seeming to rely heavily on the ritual and theology fixed by the Lipan Apaches and later taught to the Comanches and Kiowas.

Mescalism

The role of mescalism in the development of the modern peyote religion is disputed. The mescal bean comes from *Sophora secundiflora*, a plant in the pea family (Fabaceae). The bean contains a powerful and toxic alkaloid called cytisine, which produces nausea, convulsions, a numbing of the limbs, hallucinations, unconsciousness, and even death through respiratory failure.

The beans are boiled in water, and then the concoction is drunk in small quantities to produce a type of delirium, which may cause "a visionary trance" (Schultes and Hofmann 1979, 57). The beans were also used as a war medicine in the nineteenth century (Howard 1962, 130).

The mescal bean has long been used in northern Mexico and Texas, where it grows as a native plant in the Chihuahuan Desert. Use of the mescal bean raised so much concern among Roman Catholic priests in Mexico in the eighteenth century that they included it in the ritual of confession (García 1760, 15). One of the questions asked by the priest was whether the confessor had eaten *frijolillo* and if it had intoxicated him. Apparently, the ceremonial use of the mescal bean then spread northward onto the Plains, where the bean became a power fetish among the Apache, Comanche, Delaware, Iowa, Kansas, Omaha, Oto, Osage, Pawnee, Ponca, Tonkawa, and Wichita tribes (Howard 1957, 76). Native American medicine men of Coahuila, Mexico, used the mescal bean in communal dances as an intoxicant, and only secondarily as a producer of visions. However, in the Plains area of the United States, the bean came to be associated primarily with the vision-quest initiation ceremony and the obtaining of power from animals (Troike 1962, 955). There is also some evidence that it was part of the war and hunting rituals of the Caddos, Pawnees, and Wichitas.

Throughout the United States the mescal bean came to be used in several different ceremonies, but the most elaborate and important one was that of the mescal-bean society of several Plains tribes. Participants used the beans to induce visions and prolonged unconsciousness in order to gain "animal power" and to learn songs so that they could become medicine men. The bean was not treated as an object of worship but simply as the means of obtaining visions and power. Rudolph Troike (1962) described this ceremony as "a dance in which the participants decorated themselves with white paint, carried a bow and arrow and gourd rattle, and blew on long whistles to imitate the elk. Foxskins were often worn. When the dance was held as part of a first-fruits ceremony, and apparently at other times as well, a drink made from the mescal bean was taken as an emetic to purify the participants" (948).

As Native Americans of the Plains were forced onto reservations, it became more and more difficult for them to continue using the mescal bean, which was primarily a practice of a few select people who belonged to a medicine society. These medicine societies and their associated rituals soon disappeared and were forgotten. However, the memory of the vision-producing bean probably helped prepare Native Americans to accept the peyote cult when it was introduced a short time later.

The mescal-bean medicine society probably did not greatly influence the actual ceremonial procedures of the modern peyote religion, although Howard (1957, 85–86) reported that the Comanches, Otos, and Tonkawas mixed peyote and mescal beans in the making of a drink. Later these tribes stopped consuming the mescal bean but retained it as a part of the regalia worn during the peyote ceremony. The Comanches still use mescal beans as necklaces, on the fringes of leggings (as a safeguard against stepping on menstrual blood), and on peyote gourds and feathers (Carlson and Jones 1939, 537–38). The modern peyote ceremonial breakfast may have come from the so-called "first-fruits" ceremonies in Mexico, a practice of tribes that also used the mescal bean (La Barre 1957, 711).

Those who see a strong relationship between mescalism and peyotism suggest that during the nineteenth century, both practices diffused northward from Mexico, with mescalism preceding peyotism by several years. They further believe that peyote slowly replaced the more potent and dangerous mescal bean, and a distinct ceremonial ritual was adopted. Eventually an entirely new religion was created, based upon some combination of the Kiowa ceremony of the sacred stones, the old peyote complex of Mexico, and mescalism.

Merrill (1977, 60) and Stewart (1980, 307) argue, however, that there is little evidence supporting the influence of mescalism on the development of modern peyote ceremonies in the United States. Stewart even wonders "if there really was any important relationship between mescalbeans and Peyote except to use mescalbeans to decorate clothing or wear as jewelry" (307).

Native American Oppression

The Civil War formally ended a major crisis facing American democracy. The Union was preserved, and the nation could again proceed along a single political and cultural path. Unfortunately, Native Americans, with their differing cultural viewpoints and non-Christian beliefs, did not fit into the European American system. This conflict of cultures, which had actually begun more than a century earlier, led to the tragic and violent oppression of a once-proud people.

As Anglos began to migrate westward onto the open plains and prairies, they encountered Native Americans, many of whom were nomadic and were considered uncivilized. These so-called "savages" naturally resented the arrival of Anglos because the European immigrants began to kill large numbers of the bison—the basis of life for many Plains tribes—and to settle on land that was considered to be free territory given by the spirits for the use of all people. Violence and considerable bloodshed often resulted when members of these differing cultures came in contact. At first, Native Americans were able to migrate elsewhere to avoid the European immigrants and their culture, but as the frontier vanished, this eventually became impossible. Ultimately, Native Americans were forced to fight for what had formerly been theirs—and they lost.

Based on a traditional value of cultural uniformity, European American settlers and government officials believed it would be desirable for Native Americans to adopt European-based customs. Historically, the federal government has dealt with Native Americans in different ways during two distinct periods. Initially, it was decided that Native Americans themselves *were* a problem; later came an acknowledgment of, and an attempt to deal with, the problems *of* Native Americans. During the first period, though not "official" policy, thousands of indigenous people were killed, particularly in the Plains region. This was accompanied or followed by segregation, a policy of removing many Native Americans from their homelands to the Indian Territory of Oklahoma in a great mixing of tribes and cultures. During this first period, government officials reasoned that European-based culture and religion could be imposed on the tribes by requiring Anglo education and European dress, and prohibiting indigenous languages and Native American ceremonial and social practices. The persons designated to administer these programs—frontiersmen, missionaries and priests, and government bureaucrats—were often intolerant and inflexible people who had no sympathy for the Native American heritage, nor any desire to understand Native Americans as people. These enforced changes were catastrophic to Native Americans, and although Anglo intentions may have been good from a European American cultural perspective, Native Americans encountered new problems of such magnitude that the results were often tragic.

The peyote religion began to attract members of the various Plains tribes when they were being subdued and placed on these special reservations. To understand their situation, both primary and secondary changes must be considered. The first, and primary, change was that they had been conquered by European immigrants, whether in battle or by the signing of treaties. The government then assigned them to the reservations, often in areas that were completely different from where they originally had lived, and sometimes next to tribes that formerly had been bitter enemies. After the bison were destroyed and the hunting grounds fenced and plowed, these once-nomadic, independent people were forced to become agriculturalists who were dependent on European Americans for subsistence (which often was inadequate). This policy had the beneficial result of terminating intertribal hostilities, but it also produced groups of people cut off from their heritage and given no new identity for the present. In essence, Native Americans had become wards of the government without the constitutional rights of either aliens or citizens.

Native Americans were also greatly affected by several secondary changes, including the establishment Indian schools where English was the required language, and the introduction of Christianity in many, often competitive, denominational forms. Anglos also sold commercial distilled liquor—which became a serious problem almost immediately—and carried diseases such as tuberculosis, measles, and smallpox, which soon reached epidemic proportions.

Within many tribes, traditional Native American pride was gone, and in its place new concepts—and prohibitions—were substituted. The European American ban of all indigenous rites—an attempt to stamp out "pagan" practices of any kind—meant that Native Americans could no longer practice their religion, nor could shamans perform important curing ceremonies (Slotkin 1956, 50). Also, because Native American music and art had been centered in the religious aspects of their cultures, the prohibition of their religious activities meant that many also lost opportunities for aesthetic expression.

At the end of the nineteenth century, Native Americans had been effectively segregated on the reservations, although boundaries were modified several times when valuable land was needed by the dominant culture of European Americans. Over a period of several years, even some of the reservation land was taken away, leaving former hunting tribes as sedentary farmers, but on reservation lands that were largely unsuitable for farming.

During the second period, or phase, of U.S.-Native American relations, the government gradually adopted a different policy, one of *assimilation* rather than segregation. This shift of policy produced as many problems as it solved. As one example, many Native Americans found the competitive capitalistic economic system to be culturally incomprehensible. For many of them, their culture and experiences were simply inappropriate tools for coping with the dominant European American society into which they were forced to acculturate.

It is important to understand something of the religious beliefs of Plains tribes in order to see why the peyote cult appealed to so many of them during this period of chaos and tragedy. The Native Americans' world was full of spirits—some good, some bad. Many tribes believed that communication with a supreme being, or the spiritual realm, occurred through spirit-forces, which were usually plants, animals, and natural phenomena. Some Delaware spirit-forces, for example, were the sun, water, fire, tobacco, the moon, and wind (Petrullo 1934, 143). Humans were lost creatures if they did not have these spirit-forces to guide them through the difficulties of life.

These spirits were conceived of and approached in two different ways: through ceremonies and by visions. Individuals who communicated with the spirit-forces by means of the vision experience came to have "power," which they could then use for good or evil purposes. The shaman, or medicine man or woman, was the great visionary, and his or her powers normally were devoted to curing. However, some people used their power for evil purposes and became witches by causing illness and misfortune (Opler 1936, 146). Many of the Plains religious ceremonies consisted in part of "curing" or counteracting the power of a witch through the greater power of a good shaman. Shamanistic powers also included prophecy, control of the weather, the locating of lost objects, and advice on where the enemy might be found. Rivalries also arose among shamans, and often they would perform a ceremony to discover or learn about the power of someone they distrusted or disliked (Opler 1936, 146).

The vision experience was particularly important to some of the Plains tribes such as the Dakotas, who widely employed self-torture as a method of obtaining the vision and the power associated with it (Underhill 1957, 131). The ordeals took many forms and varied widely in degree of mutilation and hardship. Although in some tribes only mature men sought visions, the Winnebago tribe expected its young men to persevere in numerous fasting experiences from the age of eight or nine until they had their first intercourse with a woman (Benedict 1922, 2). For many Plains tribes, the vision quest was an integral part of the cultural pattern, and obtaining such visions was a primary method of securing the necessary power to protect oneself from evil (La Barre 1947, 298).

The great importance of the vision quest led Native Americans to seek other ways to induce visions rather than self-torture and fasting, and psychoactive plants were one alternative. Psychoactive plants used by various Native American groups included tobacco (*Nicotiana tabacum*), the mescal bean (*Sophora secundiflora*), jimson weed (*Datura stramonium*), and peyote (*Lophophora williamsii*).

After the world of the Native Americans had collapsed, the traditional social organization of many tribes was modified or gone, religious practices were forbidden, and physical discomfort was widespread because of disease, poverty, and strange foods. Reactions to the newly imposed European American culture varied from almost complete adaptation to total rejection. Those who chose to oppose the European American culture tried to appeal to the sentiments and traditional pride of Native Americans, and they attempted to create pan-indigenous movements with cultural and religious attractions for many tribes. In a sense, all of these movements attempted to reduce the cultural impact of European immigrants and to retain some of the now-subordinate cultural patterns of Native Americans. They attempted to reverse the process of acculturation to European American ways and to produce a new, truly Native American culture that would give them identity even under the new environmental circumstances.

The Ghost Dance

One of the most important of several pan-indigenous movements to appear during this period was the Ghost Dance. This religious movement arose about 1870 among the Northern Paiute tribe living in the California-Nevada border area. Their prophet Wodziwob claimed to have received a revelation that there was to be a great cataclysm in which the "white man" would be swallowed up, but the dead ancestors of followers of the Ghost Dance religion would be brought back on a big train (Lanternari 1965, 113–14). This religion was attractive for some Native Americans because it predicted the removal of European immigrants and provided a ray of hope for a cultural renaissance — a heaven on earth would be created. These miracles could be hastened by ceremonial dancing and singing.

In 1890 the Ghost Dance was inaugurated among the Sioux and other Plains tribes through the leadership of the Northern Paiute Wovoka, also known as Jack Wilson, who told his people that he, too, had experienced a vision and had learned a dance from God that they must perform to bring the dead back to life again. This new version of the earlier Paiute Ghost Dance spread rapidly and with much fervor among the Plains tribes, but the teachings of Wovoka were interpreted in different ways to achieve different ends, some with tragic results. The Sioux, for example, although advised by Wovoka to work hard and maintain peace with the European immigrants, corrupted his teachings to the extent that not only would the dancing bring back the dead and the bison—but it would also eliminate the European Americans with a great landslide. Moreover, the Ghost Dancers believed they were invulnerable to the guns of European Americans if they wore their bullet-proof "ghost shirts." A nervous U.S. Army detachment encountered a group of Sioux led by Yellow Bird, a medicine man, at Wounded Knee in 1890, and more than two hundred Sioux were killed (McKern and McKern 1970, 69). Almost as quickly as it had arisen, the Ghost Dance faded.

The Attraction of Peyotism

The Ghost Dance and the peyote religion both appealed to the nativistic desires of indigenous Americans at the end of the nineteenth century, but whereas the Ghost Dance, with its emphasis on the elimination of European immigrants and the return of ancestors, had resulted in disillusionment, peyotism provided a Native American religion that contained distinctly indigenous characteristics but at the same time demonstrated a degree of accommodation to European American culture and Christianity. According to Vincenzo Petrullo, an anthropologist who studied it extensively among the Delawares, peyotism

> contains no prophetic formula for the imminent extermination of the Whites and the return to the pre-Columbian conditions; it teaches acceptance of the new world, and makes possible an attitude of resignation in the face of the probable disappearance of the Indian groups as distinct peoples, culturally and racially, by insisting on the necessity of emancipation from mundane aspirations. The greater goal that the Indian should attempt to attain is a loftier spiritual realm which is beyond the reach of the Whites to destroy. (Petrullo 1934, 1)

The new peyote ceremony contained five basic elements that appealed to Native Americans:

1. traditional Native American religious symbols
2. magical elements, including visions
3. the pan-indigenous sentiment
4. a belief in salvation as well as an ethical code
5. certain parts of the Christian religion

Whereas the Ghost Dance had failed to bring happiness through miraculous events and militant action, the peyote religion offered peaceful conciliation and escape. To many Native Americans, the peyote religion, and other conciliatory cults such as Shakerism of the Pacific Northwest and the "Great Message" of the Iroquois, became their final link with a traditional past that was being systematically destroyed by European Americans. These religions enabled their followers to retain some sense of morale and social cohesiveness while being forced to adjust to European American culture. Vittorio Lanternari, in his book titled *Religions of the Oppressed*, commented that "Peyotism too, like the Ghost Dance, contained a messianic message; but whereas the Ghost Dance promised restoration of the past, Peyotism announced a new dispensation and a renewal of Indian culture" (1965, 81).

Native Americans thus found in peyote a means of religious expression by which they could identify themselves and obtain a degree of personal security while being faced with a bewildering variety of experiences in the European American's world. Peyotism helped to resolve the conflict between cultures through an integration of nativistic power, curing, and vision concepts with elements of Christian ideology and culture. The conflict of cultures was resolved even further through the formalization and institutionalization of both the Native American and European American concepts in the creation of a religious ceremonial practice that provides hospitality, consolation, welfare, security, and curing.

The peyote religion succeeded where the Ghost Dance failed. The latter had resulted in violence and rebellion, but peyotism advocated nonviolence, introspection, and meditation. Although distinctly Native American, it contained sufficient elements of European American culture to meet certain new conditions and to appear to be accommodative; in a sense, it could be seen as a sign of compromise, conciliation, and passive acceptance of the dominant culture (Lanternari 1965, 81).

One of the main reasons that Native Americans were attracted to the peyote religion was that it enabled them to continue their vision quests. Some anthropologists have suggested that peyote became an excellent substitute for the difficult fasting or self-torture ordeals, but others have felt that visions were only a secondary function of peyotism, while the primary function was medicinal (Schultes 1938, 704). Weston La Barre has described the relationship between vision and medicine by noting that, indeed, peyote is taken to cure ills "*because* a vision-producing plant obviously has power" (La Barre 1960, 52). He further added that " 'medicine' in reference to American Indians has by usage supernatural connotations, and that the *medicinal* virtues imputed to peyote were in fact based both on the visions it induces and on the 'power' that the Indian thus infers is in it. The problem is purely a semantic one" (54).

There is little doubt that some participants in the peyote ceremony experienced real visions, often stronger and more compelling than the earlier Plains visions. Also, visions were readily available through peyote to anyone who wished to experience them—provided they were willing to follow the often-difficult "Peyote Road." The ceremonial experience of eating peyote gave one "magical powers" which formerly had been associated only with membership in such ceremonies as the medicine dance of the Winnebagoes. Thus the peyote experience could be interpreted as a direct encounter with the spiritual realm.

The other significant attraction of the peyote religion was the Native American and pan-indigenous emphasis. The religion provided a definite sense of ethnic importance to a people in a transitional state. For those who were unable to adjust to European American culture, it became a solace because it was Native American—it had a tie to the past. It was something uniquely Native American, and European Americans could share the experience only with permission (Underhill 1952, 148). This reformative religion claimed that the Great Spirit created both "white" and "red" men and that they should go their own separate ways. But the Great Spirit communicated directly with the Native American through Peyote (Petrullo 1934, 4). Thus, certain important elements of the traditional Native American cultures were incorporated into a new religion acceptable to many tribes throughout the United States. This intertribal activity was greatly facilitated by termination of intertribal warfare and the nearness of several reservations (especially in Oklahoma). Young people from various tribes also came into contact in government and mission boarding schools, which greatly facilitated the spread of the peyote religion. Intermarriage, the ease of travel, the use of the English language as the medium of communication in place of diverse tribal tongues—as well as racial and economic discrimination—also greatly aided the diffusion of peyotism. Identity as an ethnic group, rather than a tribal one, provided a new measure of dignity and self-confidence.

The peyote religion came at an opportune time and was accepted eagerly by many Native Americans. However, approval and acceptance of this new cult were far from unanimous among either Native Americans or Anglos. One of the main Native American objections was that it was an imported religion from Mexico; some tribes, like the Navajos, considered it foreign just because it had arisen outside their own tribe. Some Native Americans also objected to peyotism because it tended to undercut the older tribal customs that still remained, especially with regard to the shaman's power. Moreover, bitter rivalries sometimes arose within single tribes over leadership; families were divided by those who wished to follow "the Peyote Way" and those who opposed it.

Opposition by European American society was much more intense and well-publicized. The basis of the persecution of peyotists was that the peyote religion was nativistic and ran counter to European American religion and policies (Lanternari 1965, 95). By the end of the nineteenth century Christian missionaries had become the primary opponents because they believed peyote was an "evil power whose special function it is to captivate and destroy the souls of the aborigines" (Petrullo 1934, 4). Anglo opponents were convinced that peyotism was incompatible with European American religion and mores, and went to great extremes to forbid it, often without the backing of specific laws relating to peyote (see chapter 9). These European immigrant opponents used all kinds of untruths concerning peyote to support their claims of its degrading effects. The policy of the Bureau of Indian Affairs towards peyote at the turn of the century was primarily shaped by the views of the missionaries, who were regarded as "a brave company of men and women who are fighting to save the American Indians from the degrading cult of Peyote worship" (Pierson 1915, 201).

The peyote religion met the needs of some but antagonized others. Like any religion, one must believe in it to understand and follow it.

The Spread of Peyotism

In 1891 James Mooney witnessed the modern peyote religion being practiced among the Kiowas in Oklahoma. Impressed by the ceremony that he attended, Mooney wrote that peyote "is regarded as the vegetable incarnation of a deity. . . . Indians regard the mescal [peyote] as a panacea in medicine, a source of inspiration, and the key which opens to them all the glories of another world" (Mooney 1896, 7–9). He found that the peyote ceremony was not held in conjunction with village dances, as with the Tarahumaras and other tribes of Mexico, but that among the Kiowas the ceremony was a voluntary ritual that involved singing, prayer, and quiet contemplation. He further commented that the physiological effects of peyote easily convinced Native Americans that it was "the incarnation of the God Spirit." Thus, by 1891, when described by Mooney, the modern peyote religion was a form of peyotism clearly distinct from the old peyote complex of Mexico. The peyote plant itself may have been brought from Mexico, but it carried little of the ceremonial ritual with it; ritual practice in the United States, as stated earlier, developed mainly from certain aspects of Native American Plains cultures and from Christianity.

The actual steps of the spread of the modern peyote religion are complex and poorly understood, but nevertheless, a number of conclusions can be drawn. Although Mooney believed the peyote religion came originally from the Lipan Apache, La Barre (1989, 122) and Stewart (1987, 46–53) argue persuasively that the religion arose first among the Carrizo of northeastern Mexico. In fact, according to Stewart, "At least as early as 1649 . . . the Carrizo were engaged in the ritual use of peyote involving an all-night ceremony, singing and drumming around a circle. The Lipan, on the other hand, were newcomers to the area after 1770" (49). Stewart concluded that from the Carrizo and Spanish missionaries, "both the Lipan and Mescalero Apache knew about peyote and Christianity by 1770—nearly a century before the earliest documented evidence of peyotism in the United States. The transmission of the peyote ceremonies were direct from Laredo by known Lipan Apache, who were named as the first teachers of peyotism to the Kiowa and the Comanche on their reservation in Oklahoma in the late 1870s and early 1880s" (1984, 271).

Thus the first U.S. tribes to receive peyotism from the Carrizo Indians of northeastern Mexico probably were the Lipan and Mescalero Apaches. By the late nineteenth century many tribes probably had received multiple contacts with peyote as intertribal contacts were more frequent, especially in areas such as Oklahoma where many tribes lived in close proximity on reservations.

By the beginning of the twentieth century most Plains tribes had been introduced to the peyote religion, the more significant ones being the Oto

(1876), Caddo (about 1880), Arapaho (1884), Delaware (1886), and Cheyenne (before 1900) (La Barre 1989, 121–23). During the first half of the twentieth century the religion spread even more vigorously; by the end of the 1970s it was found throughout the United States and in Canada, and apparently was the most widespread and popular intertribal Native American religious movement.

One of the more interesting aspects of the diffusion of peyotism in the United States has been the peyote prophets: those persons responsible for proselytizing. Some were already great tribal leaders, but others became leaders only after being converted by taking peyote and receiving a revelation. Some leaders have been accused of being opportunists who were in the business simply for material gain. Others, however, were sincere and devoted to their new religion.

Perhaps the most exciting prophet of peyotism was John Wilson of the Caddo tribe, who many claim was the "founder" of the peyote religion in the United States. This was not the same person as Wovoka, or Jack Wilson, the Northern Paiute leader of the Ghost Dance movement of 1890. John Wilson, of mixed blood which included Delaware, Caddo, and French, was introduced to peyote at about the age of forty while he was with the Comanches. Wilson decided that he would "learn what peyote might teach him" so he went to a secluded spot for a period of two to three weeks. During this time he ate peyote at frequent intervals and "Peyote took pity on him" (Speck 1933, 541). Wilson claimed that he was continually translated in spirit to the "sky realm" by Peyote, and it was there that he learned of the events in Christ's life and the relative positions of several of the spirit-forces such as the Sun, Moon, and Fire. He reported that he was shown Christ's grave, now empty, and that Peyote instructed him about the "Peyote Road," which led from Christ's grave to the Moon (this had been the "Road in the sky" that Christ traveled in his ascent). Peyote urged Wilson to travel this same "Road" for the rest of his life by obtaining knowledge through the use of peyote. Eventually, just before death, the Road would lead him into the actual presence of Christ and Peyote.

John Wilson returned from seclusion and began to conduct ceremonies as he had "learned" them during his revelation. He insisted that he was not a prophet and that *"he was not sent by God to fulfill a mission,* but that he was *shown* by Peyote how to conduct religious worship in the Peyote meetings in order to cure disease, heal injury, purge the body from effects of sin and to lead the Indians to reach the regions 'above,' *hukweyun* in Delaware, or heaven, where they would *see Peyote and the Creator*" (quoted in Speck 1933, 550). The ritual forms practiced by Wilson contained considerable Christian influence. However, he insisted that "the Bible was intended for the white man who had been guilty of the crucifixion of Christ and that the Indian who had not been a party to the deed was exempt from guilt on this score and that therefore, the Indian was to receive his religious influences directly and in person from God through the Peyote Spirit, whereas Christ was sent for this mission to the white man" (547).

Wilson also formulated a set of moral instructions, which included abstinence from liquor; restraint in sexual matters; matrimonial fidelity; and prohibitions against angry retorts, falsehoods, vindictiveness, vengeance, and fighting. He also strongly condemned the use of witchcraft and malevolent conjuring. These teachings have come to be known to peyotists as the ethics of the "Peyote Road." To purge oneself of sins was the function of peyote. Wilson insisted that the greater the amount of sin or impurity to be cleansed by peyote, the greater the amount of peyote that would have to be consumed. He added that even the amount of nausea and sickness suffered by the taker was in proportion to the amount of impurity that had to be purged. The influence of John Wilson was considerable and many of his teachings are still followed among peyotists from various tribes.

Another peyote prophet, John Rave, was converted while attending a peyote ceremony. Rave gave the following account of his conversion:

> In the middle of the night I saw God. To God living up above, our Father, I prayed. "Have mercy upon me! . . . Let me know this religion!" . . . I had been frightened during the night but now I was happy. . . . I seemed to see everything clearly. . . . O medicine, grandfather, most assuredly you are holy! All that is connected with you that I would like to know and that I would like to understand. Help me! I give myself up to you entirely! . . . Throughout all the years that I had lived on earth, I now realized that I had never known anything holy. Now for the first time, I knew it. Would that some of the Winnebagoes might also know it! (Quoted in Radin 1923, 390–91)

Rave actively proselytized on the northern Plains for a number of years with mixed success.

Some Comanches believe that the peyote religion first came through the great Comanche chief Quanah Parker, who "first learned how to use peyote in a religious way from a Mexican Indian woman who doctored him while he was very ill in the state of Texas" (Brito 1989, 3–9). In fact, his "illness" was from a deep wound caused by a Spanish bull, which led to blood poisoning. The medicine used by the woman was "woqui," a strong drink made from peyote (Stewart 1987, 72). Parker developed into an influential peyote disciple who had good relationships with European Americans, including missionaries and government officials. Though he never professed to be a Christian, he fully appreciated its good points.

The Delaware tribe illustrates another aspect of the spread of peyotism. Early attempts to introduce peyotism to the Delaware Nation failed until a tribal leader named Elk Hair became converted. He developed a strongly anti-Christian, more strictly Native American ceremony that he called the "Little Moon," in contrast to John Wilson's "Big Moon" ceremony (Petrullo 1934, 31–32). He insisted that the peyote religion should be strictly Native American and should avoid any association with European American religions. Over a period of time, however, this strongly pro–Native American variation became absorbed into the more popular ceremonial patterns containing some elements of Christianity.

In contrast to the anti-Christian leanings of Elk Hair, other peyote "missionaries" such as Jonathan Koshiway and Ben Lancaster promoted slightly Christianized and popular versions of peyotism. In fact, the Koshiway version, which had been widely accepted by the Omahas and Winnebagos, eventually became the Firstborn Church of Christ (Slotkin 1956, 58). This church preceded the organization of the Native American Church, a pan-indigenous organization that has persisted to the present.

Some anthropologists have suggested that the success or failure of peyotism within a certain tribe was primarily due to the personality of the proselytizer or prophet who introduced the cult (Stewart 1944, 97–98). Certainly this has been a factor, but it may not be the most important one. Many other aspects were involved, including the tribal leaders' attitudes and degree of control, the cultural homogeneity of the tribe, the strength of Christianity or native religion within the group, the economic situation, the opposition to peyotism by European Americans working with the tribe, the date when peyote was introduced, its route of migration, and the tribe's distance from Oklahoma, the center of diffusion. All of these variables have influenced peyotism's differential distribution. Apparently, a group's acceptance of the new religion was greater where there was neither a relatively large number of strongly traditional full-bloods nor highly acculturated mixed-bloods (Hurt 1960).

The continuation of peyotism among the indigenous people of the United States and Canada is due primarily to the Native American Church. Koshiway's Firstborn Church of Christ was incorporated in Oklahoma in 1914 following an example set in the same state in 1906 by a group calling itself the Association of Mescal Bean Eaters. This latter group later changed its name to the Union Church (Slotkin 1956, 57–58). One of the main reasons for the incorporation of these "churches" was to avoid religious persecution by European Americans.

In October 1918 several peyote groups within Oklahoma incorporated to become the Native American Church. This name arose as the result of a conflict among some of the Firstborn Church members and other peyotists over the matter of smoking. Fundamentalist Protestant and Mormon influence was strong within the Firstborn Church, and the ritual of smoking corn-husk cigarettes had been dropped. Therefore, during the conference of peyotists who were trying to work out a merger and new organization, the smoking problem and other elements led to the rejection of the name "Firstborn Church of Christ" and the adoption of the name "Native American Church," which emphasized that it was a nativistic pan-indigenous religion (La Barre 1989, 168–69). The Articles of Incorporation, signed on 10 October 1918, are quoted in Stewart (1987, 224).

> The purpose for which this corporation is formed is to foster and promote the religious belief of the several tribes of Indians in the state of Oklahoma, in the Christian religion with the practice of the Peyote Sacrament as commonly understood and used among the adherents of this religion in the several tribes of Indians in the State of Oklahoma, and to teach the Christian religion with morality, sobriety, industry, kindly charity and right living and to cultivate a spirit of self-respect and brotherly union among the members of the Native Race of Indians, including therein the various Indian tribes in the State of Oklahoma.

Within a few years, branches of the Native American Church were incorporated in other states, and in 1954 it became international with the incorporation of the Native American Church of Canada. In 1925 there were an estimated 13,345 peyotists (Newberne 1925, 35). The present number of members is estimated to be 250,000, though some people claim there may be as many as 300,000 among seventy different tribes (Cousineau 1993).

Membership in the Native American Church consists of four large organizations, each containing numerous chapters. Many state chapters do not maintain membership lists, thus making it virtually impossible to know the exact number of members in the Native American Church. According to Dayish (1994), the larger organizations are:

1. the Native American Church of North America, with 46 chapters in 24 states, Canada, and Mexico;
2. the Native American Church of Navajoland, with 92 chapters within the Navajo Nation;
3. the Native American Church of Oklahoma, with 17 chapters in the state of Oklahoma;
4. the Native American Church of South Dakota, with several chapters within the state.

Apparently these structured peyotist organizations have been effective in defeating both Anglo and Native American opponents who wish to prevent the practice of the peyote religion.

The modern peyote religion that has developed within the United States during the past century has met the needs of many Native Americans who have sought to retain important aspects of their traditional cultures, yet accommodate to the pressures of the dominant European immigrant culture. The Native American Church is now a significant institution affecting thousands of Native Americans throughout North America.

Peyote Ceremonies

The peyote ceremonies described in this chapter are unique to the Native American Church and are remarkably homogeneous among the tribes of the United States and Canada, although they differ greatly from the peyote ceremonies of Mexico (see chapter 1). This great similarity of ceremonies is due in part to the fact that the modern peyote cult of the United States and Canada is societal in its organization rather than strictly tribal; that is, membership in the peyote "church" is distinct from tribal membership and crosses tribal lines.

The modern peyote ceremony is an all-night meeting in which participants sit inside a tipi or other structure facing a fire and a crescent-shaped altar. The ceremony consists basically of four parts: praying, singing, eating peyote, and quietly contemplating. Usually those who are present participate in all parts of the long and tiring ceremony, but it seems that most of the time an individual just sits quietly looking at the fire and the "Father Peyote" — and contemplating. It is, however, a collective rite, and although in a sense the individuals are isolated from the other members in their own personal thoughts and prayers, they quickly respond when it is their turn to sing or drum. The prayers, songs, and quiet contemplation, coupled with the effects of peyote, frequently lead to personal revelations. These are often in the form of visions and audible messages directly from Peyote or the Great Spirit. Peyote often "speaks" to the participants and promises them forgiveness of their sins; members are confident that Peyote will overcome both bodily and spiritual ills, for it is the "comfort, healer, and guide of us poor Indians" (Slotkin 1956, 77).

Preparations for the Ceremony

The goal of many Native Americans is to visit the "peyote gardens," making a traditional pilgrimage to collect the peyote plants that will be used in the ceremonies (Morgan and Stewart 1984, 284). The ritual of this pilgrimage is not as elaborate as that followed by Mexican peyotists, such as the Huichols (see chapter 1). However, there are a number of rules that must be obeyed; for example, pilgrims may not eat salt, they must abstain from sexual intercourse, and they must not include any women who are menstruating (McAllester 1949, 18). The trip to southern Texas or northern Mexico may last a week or longer, depending on the elaborateness of the ritual followed and the success in collecting a sufficient supply of plants.

The pilgrims normally select a leader when they arrive in the peyote region; he or she then leads them in making cigarettes of tobacco, smoking, and praying. Peyote "never reveals itself until after the seekers have prayed, when it may appear in the form of a man or deer, leaving the plant behind" (Underhill 1952, 144). They pray to an uncut peyote plant, then place their cigarette butts in a circle and begin collecting the plants. The Comanches do not collect any plants the first day but instead hold an abbreviated ceremony in the area where the plants occur. The following day they collect only enough plants for a second ceremony that night, so it is not until the third day that they start the general collecting (McAllester 1949, 19).

Some tribes must follow certain ritual requirements connected with the actual harvesting process; a major one observed by many groups is that the top of the plant is cut off with a knife, but thereafter no knife should ever touch the "button" (the harvested top of the plant) (fig. 3.1). The roots are left in the ground so that the plant is not destroyed and can produce other tops for future use. It is not unusual to have a group collect several thousand tops, which are dried and then used by the tribe in the following months.

Most peyotists do not participate in making a pilgrimage to collect plants but simply purchase the buttons by mail or from a dealer who passes through the reservation. The first recorded trade in peyote was in the 1880s between Hispanic *peyoteros* and Native Americans (Morgan and Stewart 1984, 286). By the turn of the century an extensive peyote trade had developed in four counties of south Texas, primarily involving Hispanic families who shipped peyote by both mail and rail to various tribes throughout the country. A zealous federal government official attempted to quell this trade, destroying large

FIGURE 3.1. Peyote "buttons" are the harvested tops of the plants. They may be eaten fresh or dried (as shown here). Note the characteristic white hairs, which are often removed when the button is eaten.

quantities of harvested peyote in 1909 (Morgan and Stewart 1984, 289). Trade continued, however, with plants selling for $2.50 per thousand. By the 1940s the price of peyote had risen to $5.00 to $10.00 per thousand, and by the 1980s it had soared to $80.00 per thousand (Morgan 1983a).

There are currently eleven peyoteros operating in south Texas, and each harvests 200,000 to 300,000 heads a year. This is not enough to adequately supply the Native American Church, which has an annual demand of 5,000,000 to 10,000,000 "buttons" (Anthony Davis, pers. comm.). Peyoteros must seek permission to enter private ranches, often paying leasing fees for the right to harvest peyote. A team of five can harvest up to 30,000 heads in about five hours, using machetes or special short-handled shovels to cut the plants at, or just beneath, the surface of the ground (Salvador Johnson, pers. comm.). The remaining root system will produce one to several new stems within a few months. However, peyoteros prefer to let a population of peyote regrow for a minimum of five years.

Each peyotero is registered by the State of Texas and must keep careful records of the number of heads harvested and sold. Typically, the peyotero charges $.15 to $.17 per head, depending on current supplies and demand.

Thus, 1,000 dried "buttons" now cost $150 to $170. Calculating that each of the thirty or so participants consumes ten to fifteen buttons during a ceremony, an evening's supply of peyote for a group of might cost more than $50. After adding the expense of the meal, the total cost of a peyote meeting may now exceed $150 (Brito 1989, 78). The cost has become so great that quantities may often be limited in ceremonies. One Winnebago elder commented that in early days people "ate more medicine than we do today" (107).

A special relationship exists between peyoteros and members of the Native American Church. Peyoteros frequently assist pilgrims in going to the "peyote gardens" or wild populations when they visit south Texas. Amada Cárdenas, a revered peyotero now in her 90s, began harvesting peyote in 1933. Some Native Americans even call her "Queen Peyote." Her home and garden near Mirando City are visited by numerous Native Americans each year, where ceremonies are conducted in either a tipi or hogan located on her property.

A peyote meeting may be called for a number of reasons and by almost anyone. The usual purpose for calling a meeting is to cure an ill person, but meetings can also be called to give thanks for being cured, to pray for loved ones who are away at school or in the military, to celebrate birthdays and holidays, or to bless or help someone having a particular problem. Some tribes try to schedule meetings regularly on Saturday night, but special meetings can be called any time. Normally the person who calls the meeting also "sponsors" it; meaning that he or she obtains the peyote, arranges for the quarters and officials, and provides food for the breakfast. Other members may make the arrangements if the sponsor cannot afford the food and other items; in several tribes the Road Chief, or leader, provides the peyote.

Preparations for the all-night meeting vary widely among the tribes. Some Plains groups continue their ancient practice of having a sweat bath prior to the meeting, whereas others make no formal preparations of any kind. The ceremony usually takes place in a special tipi painted with appropriate symbols and placed on a dedicated site. Some tribes no longer use the tipi but hold their meetings in wooden buildings or even outside, with just a canvas wall around the ceremonial area. Not surprisingly, the Navajos use hogans for their meetings.

The officials are responsible for preparing the meeting site, which includes erecting the tipi or cleaning up the meeting building, collecting firewood, securing the ceremonial articles, arranging for the serving of the food, and making the altar. Some peyote groups have built concrete altars in their permanent meeting places, but usually an altar is constructed for each ceremony and then destroyed. Normally there are four officials who preside and help lead the meeting: the Roadman (also called the Road Chief, or simply "the leader"), the Chief Drummer, the Cedarman, and the Fire Chief (Stewart 1987, 346). Each has certain functions to perform during the ceremony; among most tribes an aspiring peyotist progresses from one position to another by learning the necessary functions, prayers, and songs for each office. Peyotists emphasize that the Roadman is not a priest or intermediary but rather a leader of the meeting.

Usually the Roadman supervises the construction of the altar and provides the necessary ceremonial paraphernalia. Although altar structures vary greatly, most consist of a raised, crescent-shaped mound or "Moon" with a groove running from tip to tip; this furrow represents "the Peyote Road" on which believers must travel to gain complete knowledge of peyote. The shape and height of the crescent may vary, and some tribes construct other lines and smaller mounds to go along with the basic altar. The Cheyenne-style altar, for example, usually has tips curving toward the east, and the fire is built on the floor within the area formed by the crescent (fig. 3.2). The Comanche altar is more horseshoe-shaped, because "it's supposed to represent the hoof of the donkey which carried the baby Jesus and his mother" (Brito 1989, 33).

The items of peyote paraphernalia, though somewhat variable in construction, are otherwise remarkably uniform among the different tribes practicing modern peyotism. The Roadman usually carries a special satchel or box containing the paraphernalia: the iron pot that will be made into a water drum, a piece of buckskin for the drumhead, a cord and stones with which to tie the drumhead on the pot, a drum stick, a gourd rattle, an eagle wing-bone whistle, a cluster of sage, a staff, various feathers, containers for the peyote, an altar cloth, a bucket for water, a bag of cedar incense, a bag of tobacco, corn husks or cigarette papers, and a fire stick.

Participants in peyote meetings usually are adult members of the Native American Church, but guests are welcome among most tribes. Some Plains tribes, such as the Lipan Apaches and Arapahos, forbid women, whereas others openly invite both women and children to participate in the meetings (La Barre 1989, 60). Most peyotists now wear modern clothing at the meetings. Some members, however, may paint their faces and wear native or special dress, which can include special peyote jewelry, headdresses, leggings, moccasins, buckskin dresses, and scarfs. Usually they wear these items only

Preparations for the Ceremony : 53

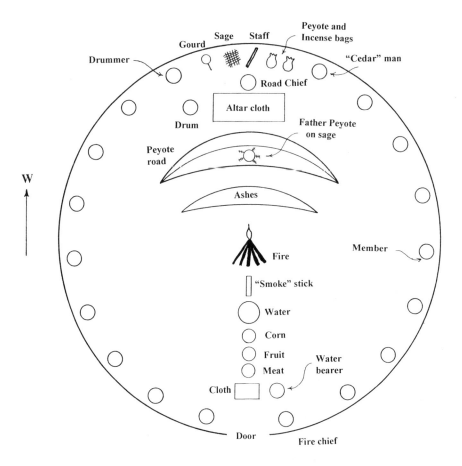

FIGURE 3.2. The basic Plains peyote ceremony is performed in a tipi or other building with a crescent-shaped altar on the floor.

to peyote meetings. Native American jewelry fashioned with a peyote theme is some of the finest made and is highly treasured.

THE BASIC PLAINS CEREMONY

The following is a description of a typical peyote ceremony of the Kiowa Comanche, also called the Half Moon ceremony, which was originated by Quanah Parker. However, each leader and "sect" has slightly different patterns and procedures that may incorporate traditional tribal customs or Christian elements to a greater or lesser extent; some of these variations are discussed later in this chapter. Nonetheless, the basic Plains ceremony described here includes most of those aspects which are common to tribes having members of the Native American Church. More detailed descriptions may be found in Stewart (1987) and La Barre (1989).

Shortly after dark, the participants enter the special peyote tipi led by the Roadman, or leader, who precedes them through the east-facing door and moves in a clockwise direction around the center altar and fire. When the Roadman stops at the west side of the tent, some of the members pass by him to the north, while others stop at the south side. Blankets usually are spread on the floor on top of sagebrush around the edge of the tipi so that participants can be seated fairly comfortably for the long ceremony. Most of them enter the tipi and find their seats informally, but some immediately kneel and begin to pray. The Chief Drummer seats himself to the Roadman's right, and the Cedarman sits to the left. The Fire Chief occupies a place to the right of the door so that he can leave easily to get firewood and carry out other necessary tasks.

After all participants are seated, the Roadman unpacks the paraphernalia. The Chief Drummer assembles the drum by placing some water and live coals in the kettle and carefully tying the wet buckskin onto the top using the small rounded stones and cord. He tests and tunes the drum by blowing air into it and wetting the buckskin by shaking the kettle.

The Roadman places a large, especially fine peyote button on some sprigs of sagebrush (*Artemisia* spp.) in the center of the crescent-shaped altar, for "peyote is friendly to no plant but this" (Shemluck 1982, 312). This is the "Father Peyote" or "Chief Peyote" to which prayers and songs are directed. This signifies that the ceremony has begun, and all informality ceases. Next the Roadman takes the bag of tobacco and corn husks, makes a cigarette, and passes the makings clockwise to his left. Each person, in turn, makes a cigarette and passes the materials on to the person on his or her left until all have received them. Meanwhile, the Fire Chief takes a special stick, called the smoke stick, and places it partly in the fire so that one end begins to burn; he takes it from the fire, blows out the flame so that it only glows, and presents it to the Roadman so that he can light his cigarette. The smoke stick is then passed clockwise around the tipi with members blowing on the glowing tip and lighting their cigarettes. When all participants have lit their cigarettes, the Roadman prays, asking the Creator, or Great Spirit, and Peyote to be with them during the ceremony and especially to be with the sponsor of the meeting who has a special need. Others may then pray aloud or silently after the

Roadman's prayers, but soon all become silent and snuff out their cigarettes on the ground. Each member rises, goes to the altar, and carefully places the cigarette butt along the west side of the crescent.

The smoking ceremony is immediately followed by an incense-blessing and the passing of a sprig of sagebrush. The Roadman, or sometimes the Cedarman, stands over the altar, takes some ground cedar from the incense bag, and sprinkles it on the fire; a pleasant-smelling white smoke quickly spreads throughout the tipi. The Roadman then takes the bag of peyote and a sprig of sagebrush and waves them in the white smoke four times to bless them. He sits down; pats his forehead, chest, shoulders, thighs, and arms with the sprig; and passes it to the person on his left who does the same. While the sagebrush sprig is passed around the tipi, the Roadman opens the bag containing the peyote buttons, takes out several pieces, and passes the bag to his left. Thus, the bag of peyote follows the sprig of sagebrush clockwise around the tipi until everyone has rubbed the sagebrush on his or her body and has taken up to four peyote buttons.

All members then begin to eat the peyote, first plucking out the wool from the top if it has not been removed prior to the meeting. Some chew the button for a few moments, spit it into their hands, roll it into a ball, and quickly swallow it, partly to avoid the strong bitter taste. Many even rub their hands and arms with the peyote juice remaining on their hands, for peyote is "powerful medicine." More peyote can be called for at any time during the night, and if participants become ill from eating the plant during the ceremony, they are permitted to leave the tipi until they feel better.

When all have eaten their peyote, the Cedarman throws more cedar incense on the fire, and the Roadman blesses the staff, feathers, and gourd rattle in the smoke. He takes another sprig of sagebrush, the staff, and feathers in his left hand, and he holds the gourd in his right. The Chief Drummer takes the drumstick, blesses it in the smoke, and begins to beat the drum. The Roadman starts to shake the gourd and sings the Opening Song and three others. When he has finished singing, he passes the paraphernalia to the Chief Drummer, who exchanges it for the drum; the Roadman then drums while the drummer sings a group of four songs. Next the paraphernalia is passed to the Cedarman, and while the Chief Drummer or Roadman drums for him, he, too, sings a group of four peyote songs. After those three officers have sung, the drum and paraphernalia pass clockwise around the tipi so that each person has an opportunity to sing a group of four peyote songs. As the drumming and singing continues, others usually sit quietly, staring at the fire and "Father Peyote," or pray aloud in a low voice to Peyote. "While you are singing, look at that Chief Peyote, and it will help you reach God" (Brito 1989, 72). The participants continue to sing and pray until midnight, with the Fire Chief carefully keeping the fire burning. Many feel that the flames are God talking to them (Snake 1993).

At midnight the ceremony changes. Either the Cedarman or the Roadman puts incense on the fire and calls for everyone to bless themselves in the smoke. The Fire Chief takes the eagle wing-bone whistle and goes outside, where he blows the whistle four times in the direction of each of the four cardinal points. He then returns for the singing of the Midnight Water Song by the Roadman, which is repeated four times. The Fire Chief again leaves the meeting place and soon returns with a bucket of water, setting it east of the fire; he kneels behind it on a blanket and faces the altar. The Roadman again blows four times on the eagle wing-bone whistle, drawing it twice through the water that is in the bucket to bless it. He carefully places the altar cloth west of the altar and arranges the paraphernalia upon it. The Fire Chief, while kneeling on the blanket east of the water and fire, is passed corn husks and tobacco, makes a cigarette, puffs on it four times, and prays. He passes the cigarette to the Roadman, Chief Drummer, and Cedarman; each puffs four times and prays in turn for anyone who is sick or in need of help. The butt is placed at the altar with the others, and then the bucket of water and a cup is passed so that members may drink. Informal conversation is permitted while the bucket is passed, and the Fire Chief carefully sweeps the area around the altar, adds wood to the fire, and builds up more ashes on the ash moon he is making on the east side of the altar. The bucket is removed from the tipi after all have had a drink. Then the Roadman picks up the paraphernalia, and the singing and praying begin again. Peyote may be eaten at any time. Participants are permitted to leave the tipi to stretch and relieve themselves after the midnight ceremony.

At the first light of dawn, the Roadman instructs the Fire Chief to get the woman who is to bring another bucket of water. When she arrives, the Roadman blows the whistle four times; she enters and places the bucket of water east of the fire, and like the Fire Chief at midnight, sits or kneels on a blanket just east of it. The Roadman collects the paraphernalia from those who had been singing and sings the four Morning Songs. Just as at midnight, he then spreads the altar cloth and arranges the paraphernalia on it. The woman proceeds to make a cigarette, smoke it, pray, and pass it on to the officials, who also puff from it and pray. At this point in the meeting the Roadman or some other designated person may perform a curing ceremony for any patients present. After doctoring is completed, the water is passed, and all again drink communally from the bucket with a ladle or cup.

In some tribes a ritual breakfast of meat, fruit, and maize are passed into the tipi with the water when the woman enters, whereas other groups wait until all have drunk the water before having the food passed in. If the water has been drunk before the food is passed in, then the leader of the meeting sings a group of four final songs, the last of which is called the Closing or Quitting Song. Otherwise, these are sung before the food and water are passed. The bowls containing the food are placed in a specific order from west to east: water, maize, fruit, and meat. When the food is properly arranged and the singing completed, the Chief Drummer takes apart the drum. The base of the drum and the other paraphernalia are passed around clockwise from hand to hand, and all sip a little of the water from the drum base as it comes to them. The Roadman takes the "Father Peyote" from its place on the altar and puts it and all the other paraphernalia away.

The formal part of the ceremony is over, and quickly everyone relaxes, smiles, chats, and sings. The food and water are passed around the tipi so that all participants may take as much as they want in their fingers or hands. When the ritual breakfast is finished, the Roadman instructs the Fire Chief to lead the participants out of the tipi. The group files out, the fire is extinguished, the altar removed, and the tipi taken down.

Later in the morning the sponsor of the meeting provides a large feast for the participants and their families so that they may be full and rested before leaving for home later in the day.

Variations of Peyote Ceremonies

Some interesting variations of the basic Plains ceremony have been reported in the extensive literature since the beginning of the twentieth century. Nevertheless, despite the following relatively minor differences, the ceremony has remained remarkably similar throughout its geographical range.

Different Types of Peyote: One of the most interesting variations is that of the Carrizo and Lipan Apaches, for they recognize two kinds of peyote: the "male," which has red flowers, and the "female," which has white (M. E. Opler 1938, 279). The Iowas do not recognize different sexes of peyote plants, but during a meeting, they do distribute peyote differently to men and women: males receive either two, four, six, or eight buttons, whereas females always get two at a time (Skinner 1915, 725).

Anthony Davis (pers. comm.), a Pawnee Road Chief also known as White Thunder, said he may eat up to fifty buttons at a meeting, but most participants eat twenty or fewer. Davis believes that a seven-ribbed plant is the most valuable, whereas others prefer those with twelve or thirteen.

Most members of the Native American Church prefer to eat "green" or fresh peyote, but this is often not possible because of the time and distances involved in transporting the buttons from south Texas to various meeting places in the United States and Canada. Most church members are unable to go personally to the "peyote gardens" of Texas, but they can get them from peyoteros in south Texas who dry and ship buttons through an active mail-order business. The dried buttons are inexpensive to ship and will keep indefinitely.

Peyote is used quite differently by the Cheyennes. In addition to eating buttons and drinking peyote tea, they also rub the juice from peyote (usually mixed with saliva because it has been chewed first) over the body as a sign of its external therapeutic powers (Hoebel 1949, 128). The peyote prophet John Rave also used peyote tea in his rite of baptism. He dipped his fingers in the tea and passed them over the forehead of the new member in a manner similar to the Christian clergyman who uses water (Radin 1914, 3–4).

Restrictions on Participation: Although many tribes allow both sexes and all ages to attend meetings, others, such as the Tonkawas (M. E. Opler 1939, 434), Apaches (M. E. Opler 1938, 278), and Arapahos (Underhill 1952, 145), do not allow women to attend. The Washo and Northern Paiute tribes allow women to attend, but during menstruation they must tie feathers on their wrists (Stewart 1944, 78).

Vomiting: People sometimes become ill from eating peyote, so a number of tribes have made provisions in their ritual to allow participants to bring tin cans so that they will not interrupt the meeting by needing to leave the tipi (Malouf 1942, 99). Nearly all peyotists emphasize that "vomiting cleanses the entire system of its filth, purifies the blood, and restores a person's nervous system to its normal condition" (Petrullo 1934, 71).

Smoking: Most tribes consider the making and smoking of cigarettes an important part of the Plains peyote rite. As described earlier, there is the "main smoke" early in the ceremony in which all participate in making and smoking cigarettes. Three other points of the rite also require someone to "take smoke" (Brito 1989, 68): when the Fire Chief prays over the water at Midnight, when the woman who has brought in the Morning Water prays over it, and when a participant prays over the morning breakfast. Brito (1989) further comments that any "participating member can also ask for smoke at any point in a meeting, as long as his request does not interfere with one of the four required smokes" (69).

Several U.S. tribes do not practice the traditional Plains "smoke rite." A Winnebago elder said that

> maybe one of the reasons the Winnebago peyote people gave up the "smoke way" was that in the old days, tobacco was used by our people in warfare against other Indian people from different tribes.... From one side they were under the pressure from Christian influences to give up the old Winnebago religion, and from the other side, they were troubled by fighting with other Indians. So in order to have peace, to have no more trouble with both of these outside people, many Winnebago people gave up the use of tobacco as it was used against other people in the old religion. And after that time, around 1909, when the "smoke way" came with the peyote religion to us, it just never seemed to catch on.
>
> Another reason we probably gave up the use of tobacco is that those people [Christian missionaries] told us that smoking was no good for us. (Quoted in Brito 1989, 101–2)

In contrast to most peyotist smoke groups, the Mescalero Apaches use a red stone pipe rather than cigarettes (M. E. Opler 1936, 162).

Firewood: Each Fire Chief knows the specific type of wood to burn to produce different ashes, coals, and colors. Some may even mix the types of wood to produce a variety of colors at once (Anthony Davis, pers. comm.). However, the wood is always placed into the fire with the growth end first.

Breakfast: The usual elements of the ritual breakfast (served in order of importance) are water, maize (corn), fruit, and meat, but some tribes add crackers and candy or use substitutes. The Pawnees, for example, serve only water, maize, and candy (Murie 1914, 637), while the Washo and Northern Paiute tribes use boiled rice instead of maize (Stewart 1944, 101). Salt is not taken with the meat, so some tribes do not use pork because "it is considered a salty meat" (Brito 1989, 86).

Dancing: Peyote meetings normally are quiet, sedate ceremonies involving much formal ritual with singing and praying, but it has been reported that sometimes the Washo and Northern Paiute people engage in informal and spontaneous dancing during the meetings (Stewart 1944, 80).

Sweeping: Some ceremonies involve ritualistic sweeping by the Fire Chief prior to the passing of the Midnight Water. Debris on the floor of the tipi is carefully swept outside, thus sweeping "out the bad spirits someone inside might have thrown off and left in front of him" (Brito 1989, 46). It also provides a "clean path" for the Midnight Water.

Midnight Water: An interesting variation in the passing of the water bucket at midnight is illustrated by Gosiute peyotism; these peyotists purposely pour some of the water on the ground in the tipi "to keep mother earth from drying up" (Malouf 1942, 98).

The Navajo "Double-Meeting": Roland M. Wagner (1975, 166–70) describes an interesting variation of the traditional peyote ceremony in which the altar includes a large central star upon which the Father Peyote is placed. The ceremony follows the typical peyote rite prior to midnight, but afterwards there is an "abrupt transition" into a traditional Navajo ceremony with the purpose of divination and the exorcism of witchcraft. Wagner (1975) comments that this "double-meeting" is "an obvious solution to the problem of which religion to resort to in a time of difficulty in that both are synthesized into a single ritual" (179).

The above variations of the basic peyote ceremony are merely a sample of some of the forms the peyote ritual has taken within the United States and

Canada. However, the basic ceremony involving the sacramental or religious eating of peyote remains intact; variations deal primarily with matters peripheral to the main ceremony itself and often have arisen through individual or tribal interpretations of what has been seen or heard. Peyotism has passed from tribe to tribe almost totally by demonstration and word-of-mouth; there are few written instructions regarding the leading of meetings, so it is an easy matter for Roadmen to create their own distinctive rituals, which, in their opinions, are just as authentic as any other.

Visions

One of the major reasons for eating peyote is to induce visions, so it is important to evaluate the place of visions in the modern peyote ceremony. As stated in chapter 2, visions have great significance to most Plains Native Americans, for they "provide foreknowledge and defensive powers" (Brito 1989, 168). Visions are also necessary in some tribes if one aspires to become a successful medicine man or woman. Interestingly, most peyotists claim that they do not wish to have visions during the meetings unless they result in a "revelationary experience," which, in contrast to the often-terrifying or unpleasant vision, is thrilling and rewarding. The great peyote prophet John Wilson emphasized that if one concentrates on the fire and Father Peyote while eating the peyote, one will not get sick or see visions. Visions, he claimed, are signs of bad self-adjustment; they are the result of sins and impurities (quoted in Speck 1933, 545). Ruth Shonle (1925), an American sociologist, commented that the moment John Rave "accepted the peyote as holy medicine . . . his fear left him and his visions changed from those of fear to those associated with familiar medicines" (72). It would seem that the effects of peyote are not lost, nor do they diminish for experienced peyotists; rather, individuals claim they experience inspiration, protection, power, prophecy, and salvation. For peyotists, the physiological effects of peyote seem to satisfy their emotional needs and desires, and provide a powerful religious experience.

Christian Influences

The Christian religion had little influence on the early peyote ceremonies of Mexico; however, by the time European immigrants became aware of peyotism in the United States, a number of Christian elements had been incorporated. The peyote religion apparently changed quickly from a religion of strictly Native American elements to one basically indigenous, but with "a veneer of Christianity" (see chapter 2). These syncretistic aspects are readily seen in the fact that the Native American deity is equated with the Christian God, Peyote or the Peyote Spirit with Jesus, and the messenger spirits (usually in the form of birds) with the Christian dove. Moreover the ethical concepts embraced by peyotists (called "the Peyote Road") include both traditional Native American ideals and certain fundamentalist Christian rules such as temperance, brotherly love, and care of the family.

In adopting certain Christian elements into their religion, peyotists from many tribes, facing persecution by Christian missionaries and government officials, argued that peyotism was simply one of the innumerable variants of Christianity, but one in which peyote, rather than bread and wine, was the sacrament (La Barre 1989, 163–66). They offered as evidence the fact that they belonged to a "church," that they followed the Ten Commandments and the Golden Rule, that they observed the sacraments, and that they used the Bible in some of their ritual. In fact, like members of certain fundamentalist Christian sects, peyotists are able to quote verses from the Bible as justification of their beliefs and practices. In Exodus 12:8, for example, the Old Testament writer refers to "bitter herbs"; the peyotist claims that peyote is the bitter herb. Romans 14:1–3 is used as a further justification for eating peyote and, in the opinion of the peyotist, is a biblical injunction for anti-peyotists not to condemn the eating of peyote: It says, in part, that "those who eat must not despise those who abstain, and those who abstain must not pass judgment on those who eat; for God has welcomed them." Members of the Native American Church learned the use of such proof-texts from Christian missionaries and have simply applied the same technique to "prove" the biblical validity of their religion. In fact, in some ceremonies, the Chief Peyote is placed on an open Bible (Stewart 1987, 152).

Peyote symbolism, present in both the ritual and paraphernalia, also demonstrates the Christian influence. The Menominis, for example, state that the poles of their tipi represent Jesus and His disciples; they form the ashes from the fire in the shape of a dove, carve crosses on the staff, and use the sign of the cross during the ceremony (Spindler 1952, 152). The southern Utes refer to the staff as "Christ's staff," and the twelve feathers on it symbolize the twelve apostles (M. K. Opler 1940, 472). Some claim that the three leaders of the ceremony represent the Trinity: God, Jesus, and the Holy Spirit (Stewart

1987, 160). The Arapahos claim that when they blow the eagle wing-bone whistle to the four cardinal points at midnight, they are announcing "the birth of Christ to all the world"; even the Midnight Water takes on Christian symbolism because the "reason for drinking water at midnight is because Christ was born at midnight and because of the good tidings that he brought to the earth, for water is one of the best things in life and Christ is the savior of mankind" (Radin 1923, 417). Some of John or Moon Head Wilson's Big Moon ceremonial elements among the Delaware and Plains tribes were strongly influenced by Christian elements, particularly the shape and construction of the altar and fireplace. Many Native Americans speak of this as the "Moonhead" or "cross-fireplace" (Petrullo 1934, 81–99; Brito 1989, 112, 123).

Most peyotists strongly affirm the Christian elements as an important part of their religion. One of the most interesting claims is that "Peyote was sent to the Indians and that afterwards Jesus was sent to the Whites, with the same purpose. However, the Whites killed Jesus in their ignorance, and thus have only the cross left; whereas the Indians never killed Peyote, with the result that they still have him, and the material manifestation of Peyote is the plant" (Petrullo 1934, 142). Members of the Native American Church believe that Jesus, like Peyote, is a spirit-force, so Christian immigrants and Native Americans worship the same God but just have *different* spirit-forces. Also of great significance for some Native Americans is the fact that their spirit-force is represented by a plant: "Peyote is the Indians' Christ. You white people needed a man to show you the way, but we Indians have always been friends with the plants and have understood them. So to us peyote came. And not to the whites" (Underhill 1952, 143).

Many other peyotists, however, strongly affirm that they worship the "true living God, who is Jesus Christ"; Iowas are even baptized in the name of the Christian Trinity (Skinner 1915, 726). As mentioned above, John Rave, the Winnebago peyote leader, baptized "by dipping his hand in a dilute infusion of peyote and rubbing it across the forehead of a new member, saying 'I baptize thee in the name of God, the Son, and the Holy Ghost, which is called God's Holiness'" (Radin 1923, 396). The peyotist firmly believes that he or she is in direct communication with Jesus and/or God through Peyote: "The white man goes into a church and talks about Jesus, but the Indian goes into a tepee and talks to Jesus" (La Barre 1964, 98).

Peyote Music

Peyote songs are unique and highly significant to the peyotist. A song is like a prayer and has much power; therefore, Native Americans believe that one should not sing a song at the wrong time nor talk lightly about it. These songs are the gift of Peyote and should be revered. Willard Rhodes (1958, 48) emphasized that songs are easily diffused from one group to another and that probably the peyote songs "figured importantly" in the spread of the whole peyote ritual complex from tribe to tribe.

One fascinating aspect of many peyote songs is that they contain numerous meaningless syllables. Such a syllable generally consists of one of the following consonants and certain vowels: *y, w, h, c, k, t, x,* and *n*. These syllables are then grouped into sequences that resemble words, as, for example, the important peyote "word" *heyowicinayo*. Bruno Nettl (1953) suggested that "the peyote syllabic sequences possibly derive from originally meaningful words in the language of one of the tribes from which the cult spread" (161–62), but the present forms of the songs make it very difficult, or impossible, to ascertain the actual source. Ruth Underhill (1952, 146) felt that even though many meaningless syllables exist in a peyote song, there are still enough intelligible words to serve as keys to a thought or idea so that the singer may more easily remember a song and use it at the appropriate times in the peyote meeting.

David P. McAllester (1949, 85–88), in a definitive study of peyote music, stated that apparently the style of peyote songs is derived from several sources: (1) Native American musical styles from the South and West: Music from these regions is simple, though of varied tempo and with a set formula for opening and closing each song. These tribes traditionally sang sets of four songs accompanied by a pot drum and gourd rattle; (2) The Ghost Dance music: This music consisted primarily of paired phrases with a limited range in which there was an "initial arc" in the melodic line; (3) European-immigrant influence: The rhythmic pattern with paired phrases is typical of Gospel hymns, and it seems apparent that these Protestant songs, as well as the vocal technique employed, provided an important element that was incorporated into the peyote songs. The older Winnebago songs of communication, for example, are quite irregular, whereas most of their peyote songs have a definite rhythmic pattern with simple count divisions in two-four time and a metronome speed of 130 to 150.

These influences have led to peyote songs that are distinctly Native American, especially with respect to uneven time divisions and the drum and gourd rattle accompaniment. From this basic form, peyotists have created many types of peyote songs; some tell stories, while others refer to the dawn, to birds or animals, to peyote, to Jesus, to water, or to the participants themselves, urging them to some kind of action such as repentance or prayer. The songs play such a significant part in the peyote ceremony that many Native Americans insist that they are "praying with their songs," for they come as a direct revelation from the spiritual realm, the gifts of Peyote.

Memories of a Navajo V-Way Ceremony

During the second half of the twentieth century, peyotism increased in popularity among the Navajo people due to several factors. These are discussed in an excellent and definitive book by David F. Aberle (1991), who spent many years studying peyotism and its variations among the Navajo. I had the opportunity to witness a peyote ceremony firsthand when I was invited to attend a Navajo peyote meeting on the reservation near Lupton, Arizona. I had made friends with several Navajo men somewhat earlier, and I was both surprised and pleased to learn that one of them often attended peyote meetings. I told him of my long interest in the cactus, and he in turn invited me to attend a meeting near his home. The following narrative relates my experiences on that most remarkable night.

The chilling breeze of late spring greeted us as we got out of the car that had brought us to the ceremonial hogan in eastern Arizona. My host, and the sponsor of the peyote meeting, was a student at Northern Arizona University, and he had asked me and my brother-in-law to attend the meeting. The purpose of the meeting—to pray for the host's success at the university—was somewhat unusual. Most Navajo meetings center around the need to pray for someone who is ill, or for personal guidance in business or private life. At first, the Navajo leaders had been suspicious of two Anglos being invited to the ceremony, but the young man assured them we wanted to attend the meeting to offer our support and prayers for his success. This explanation and warm handshakes turned their caution and distrust into open friendliness and hospitality.

The ceremony was to be held in a modern Navajo hogan, hexagonal in shape and constructed of horizontally laid logs with mud between. The roof, however, was not of mud, but of green composition material. Some nearby buildings and cars indicated that a family or two lived in this somewhat remote area. Just outside the east-facing door of the hogan was a rack containing firewood and a brightly burning fire in a shallow pit. Several small groups of people stood chatting around the fire. They carefully observed us as we approached; a few nodded a cautious greeting, but they hesitated to shake our hands. We had received permission to attend the meeting, but many of the Navajos still seemed suspicious of our motives. Were we there just to get "kicks" from the peyote? Had we come to ridicule them? We felt that we were being tolerated—but only grudgingly.

About 9:00 P.M. we were instructed to enter the hogan. We followed our host into the interior of the building, lit by a single bare lightbulb. The inside of the hogan was finished with sheet rock and was painted green; the floor was of brown concrete. There was no furniture in the room, but mattresses had been placed on the floor around the circular wall, on which hung a few pictures, a shelf, an electric wall fan near the single door, and a blanket. One prominent picture was a Roman Catholic portrait of Jesus Christ with a bleeding heart; another was of the Last Supper. The shelf held some family pictures and other unidentifiable objects, which apparently had nothing to do with the ceremony. In the center of the room was a dish, an enameled pot, and the altar: a low, bare rectangle of packed sand two to three centimeters (or one inch) high and about a meter (or yard) square (fig. 3.3).

We walked single file in a clockwise direction around the altar area to our places on the north side of the hogan and sat down on the mattresses. About thirty people were present, including six women and two young girls. Both men and women wore typical Navajo attire: Levi's and cowboy shirts for the men, and colorful velvet blouses and long skirts for the women. Some wore boots and others had on tennis shoes or oxfords. No one wore special clothing, although several had lovely Navajo necklaces, belts, and bracelets.

The Roadman, or Road Chief, brought a tall, distinguished-looking man over to where we were sitting and introduced us. He was a visiting Kiowa who was a great peyote leader in his tribe and in the Native American Church of Oklahoma; he would serve as Cedarman and interpreter so that we could understand what was happening during the ceremony, because much of it would be conducted in the Navajo language. He shook our hands, smiled, and welcomed us in excellent English. Four of us, two Anglo men, a Laguna woman, and a Navajo university student from another part of the reservation,

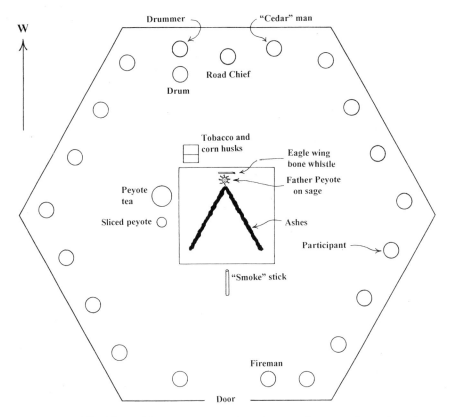

FIGURE 3.3. The Navajo V-way ceremony takes place in a hogan with a V-shaped arrangement of hot coals and ashes on the floor, forming the altar.

had never attended a peyote ceremony, so the Kiowa leader said he would explain various parts of the ceremony as the evening progressed.

When all participants were seated, the Roadman stated the purpose of the meeting in Navajo (to pray for the success of our host at the university); the Cedarman translated the Roadman's speech and told us to go outside to vomit if we became ill from eating peyote. To get sick in the hogan is unsanitary, he said, for children eat and play on the floor.

The four officers who were to lead the ceremony began to prepare the altar and paraphernalia. Unlike the typical Plains ceremony in which there is a large crescent-shaped altar and a fire within the tipi or room, this ceremony was to be the Navajo V-way ceremony, a rite having a much less elaborate altar and only hot coals and ashes rather than a fire within the room. This variant ceremony also stresses personal confession of one's sins to a greater extent than the Moon rites of the Plains. Apparently this ceremony arose during World War II; the V signifies the "V for Victory" that was such an important motto for many Navajos who were in the military service during the war. Other Navajos, however, claim that the V symbolizes Christ's victory over death. According to our host and the Cedarman, each Roadman has his own ceremony; that is, each leader has certain unique and distinctive features which make his ritual pattern different from any other within the Navajo or other tribes.

The Roadman's first formal act was to rise from his seat to place four pieces of sagebrush on the west side of the altar in the form of a small cross, about four centimeters (or an inch and a half) across. Upon this cross he placed the "Chief or Father Peyote," a large, nicely shaped fresh top of a peyote plant. The peyote drum was already assembled; it consisted of a large brass four-legged kettle, about half full of water, and a wet rawhide top bound to the kettle by a cord and stones. The other paraphernalia, taken out of a special box, included the gourd rattle, feathers, sagebrush, a bag of cedar incense, an eagle wing-bone whistle, bags of Bull Durham tobacco, cigarette papers and corn husks, an altar cloth, a staff, and a smoke stick.

When all of the paraphernalia was properly placed and the "Father Peyote" rested on the sagebrush cross upon the altar, the Fire Chief left his place at the north side of the door, went outside, and returned in a few moments with a small, flat coal scoop full of hot coals; he carefully arranged these coals in the form of a V with its point toward the Roadman, or west. After the Fire Chief returned to his seat, the Roadman stood up, threw a pinch of cedar incense on the hot coals, took a bundle of sagebrush, and blessed it in the smoke. He gave the sagebrush to the Cedarman, who smelled it; tapped his forehead, face, arms, hands, chest, and wrists; and then passed it on to the next person, who did the same.

The next portion of the ceremony was long and somewhat complex: the making and smoking of corn-husk cigarettes. The smoke stick, a round stick about twenty centimeters (eight inches) long and two to three centimeters (an inch) in diameter, had been placed next to the altar on the east side. As the Fire Chief lit the smoke stick in the hot coals, the Roadman passed corn husks (and cigarette paper) and a bag of Bull Durham tobacco so that each of us could roll a cigarette. The Fire Chief brought the smoke stick to the Roadman, who carefully lit his cigarette, then passed the glowing stick to the Chief Drummer sitting on his right. When the Chief Drummer's cigarette was lit, he handed the smoke stick past the Roadman to the Cedarman, who was on the Roadman's left. He, too, lit his cigarette, took several deep puffs, and passed the stick on to the person on his left; slowly the smoke stick was passed in a clockwise direction around the room so that all cigarettes were lit from it.

The purpose of the smoking ceremony is to bring unity to the participants as smoke from each of their cigarettes mingles in the room. The Kiowa explained to us that the use of cigarettes came from two older Native American customs. First, the Native American peace pipe was a traditional ceremony of friendship. Second, many of those in military service during World War II found that smoking cigarettes in a group, large or small, was a form of comradeship. Needless to say, we Anglo nonsmokers had difficulty properly rolling the cigarettes and smoking them, but eventually everyone finished their cigarettes. The Fire Chief then collected the butts and carefully placed them on end along the west edge of the altar, making a nearly continuous row.

Next the Fire Chief uncovered a large enameled pot containing the peyote tea and took it to the Road Chief, who drank two cupfuls from an enameled cup. He passed the cup and pot to the Chief Drummer, and then to the Cedarman, each drinking in the same manner. The Cedarman handed the pot and cup to the person on his left, and slowly the containers passed clockwise around the hogan, with each person drinking some of the peyote tea. The cup was passed from hand to hand, but the large eight-quart pot was usually slid along the floor because of its weight. The peyote tea was a brownish liquid made by boiling pieces of peyote in water and then letting it sit for several hours. Pieces of peyote floated on the surface, and some participants purposely took several small pieces in the cup with the liquid. The tea tasted strong and extremely bitter, and it was difficult to swallow. No one drank more than two cupfuls. The bowl of fresh peyote slices was taken from beside the altar and passed after the tea; most of the pieces were about an inch in diameter and consisted of tops, lower parts of the stem, and upper parts of the roots. Each participant took a single piece or "button," put it in his or her mouth, thoroughly chewed it for several minutes, and then swallowed it. Because I wanted to be able to record the events of the night, I consumed only a very small amount.

Singing began while the peyote was being passed and eaten. The Roadman began by singing a series of four songs, including the Opening Song. He was accompanied by the Chief Drummer, who was kneeling with the drum between his knees and beating a rapid rhythm with the drumstick.

The Roadman held the gourd rattle in his right hand, and in his left hand he held the staff, two clusters of eagle feathers, and the bunch of sagebrush. The Cedarman informed us that many of the peyote songs were derived from traditional Navajo medicine songs. When the Roadman finished singing, he traded implements with the Chief Drummer so that he could sing four songs. The Cedarman sang next, and then the paraphernalia proceeded clockwise around the hogan so that each person, including the women but not the guests or children, could have the opportunity to sing. Sometimes other persons beat the drum so that the Chief Drummer could return to his seat and rest. Singing continued uninterrupted for about two hours while the Fire Chief occasionally added hot coals and cared for the ash V on the altar with his coal scoop and a flat board. The Cedarman or Roadman also periodically dropped cedar incense on the coals so that the sweet-smelling smoke continued to permeate the warm and somewhat stuffy room.

Several times prior to midnight, latecomers entered the hogan, but the pair of participants engaged in singing and drumming did not stop. The newcomers made cigarettes, smoked them, and had the butts added to the row on the west edge of the altar. Then they received peyote, eating and drinking as much or as little as they wished.

After each person had sung a set of songs, the peyote was passed again. By this time all the tea had been drunk, so the participants ate pieces of peyote from the dish or that had been floating on the tea in the pot. If people got tops having clusters of hair or wool, they carefully plucked them before putting the "buttons" in their mouths. The Roadman now offered special prayers and supplications for the guests and sponsor. He tapped each of us with eagle feathers that had been passed through the smoke and gave us some large peyote buttons to eat. All of this was a special gesture of friendship. Again the singing started, with the drum and other implements going from person to person, clockwise around the room.

A little after midnight the Fire Chief brought in water in an enameled bucket with an eagle painted on the outside. He placed the bucket on the floor east of the altar and knelt behind it. The Roadman gave him materials for making a cigarette, prayed, and sang a group of songs which included the Midnight Water Song. Meanwhile, the Fire Chief rolled a cigarette, smoked it, prayed aloud, and then added the butt to those already at the altar. While the Fire Chief smoked and prayed, the Road Chief carefully placed all the paraphernalia on an altar cloth at the northwest corner of the altar. The Fire

Chief then got up, took a broom and small dust pan, and carefully swept the floor around the altar, removing any dirt and the hairs from the peyote tops. He carried the debris outside and soon returned to pick up the water bucket and start it in a clockwise direction around the room so that each person could have a drink. The Kiowa Cedarman told us that this period of the ceremony was informal and that all were urged to speak about any personal problems or to express thanks for prayers that had been answered. I expressed my appreciation to my Native American friends for permitting me to participate in the ceremony, and reaffirmed my support of the sponsor in his university studies. Several people grunted after I finished talking, indicating that they approved of and appreciated what I said. Persons also grunt during the ceremony after a particularly moving or beautiful song or prayer.

While people refreshed themselves with water, the Kiowa Cedarman talked to us informally about the peyote ceremony and its meaning to Native Americans. Quite suddenly he began to pray in English, asking that the Almighty be considerate of the sinning persons present, imploring His forgiveness, and seeking His guidance. He became more and more emotional as he prayed and soon began to weep openly; others in the hogan also wept and sobbed as the force of his prayer reached them too. The seriousness of those who participated was very impressive, and it was obviously a meaningful and valid religious experience to them. It was *not* a dangerous or wild orgy in which there was intoxication, sexual license, and other immoral activities. Peyotism clearly demands much of its followers: physical endurance, patience, confession, and repentance.

When the Kiowa peyote leader finished his prayer, he wiped his eyes and sat down. The room was very quiet except for a few sniffles and the blowing of noses. Slowly the Roadman stood and began to pray in Navajo. His words came slowly, but as he continued, one could feel the intensity of his prayer increase, and soon he too began to cry. As tears streamed down his face and words sometimes caught in his throat, there arose a chorus of grunts, moans, and sobs from the participants in the hogan. He prayed for about fifteen minutes, then stopped, stood quietly for a moment with head bowed, and sat down. Again the emotionally aroused group noisily began to relax and collect its breath. After a few moments the Fire Chief arose, picked up his flat board and a short-handled shovel, went to the altar, and knelt to the east of it. With deliberate motions he slowly and artistically changed the V-shaped pile of coals and ashes into the outline of a Peyote Bird with its head pointing to the right and its body covering most of the flat altar. He added more hot coals from the fire still burning outside, returned the tools to their places by the door, and proceeded to move all of the cigarette butts from the west edge of the altar to two small piles at the northeast and northwest corners of the altar. The Roadman arose, picked up the eagle wing-bone whistle that had been lying on the altar near the "Father Peyote," and left the hogan. For a few minutes we waited, some silently and others quietly chatting. Then we heard the shrill sound of the whistle from outside, once, twice, three times, and finally a fourth time, to each of the four cardinal points. Soon he returned, walked to the west of the altar, picked up the paraphernalia, and accompanied by the Chief Drummer, began to sing again. As before midnight, the drum, feathers, sagebrush, and staff slowly made a clockwise circuit of the hogan, with each participant singing four peyote songs.

Between groups of songs, a few people at a time were permitted to leave the hogan to stretch and relieve themselves. We went outside, walked around in the dark, breathed deeply the cool fresh air, and then stood stiffly around the fire burning near the entrance to the hogan. When four songs had been sung and the paraphernalia were being passed to the next pair of people, we again returned to our places inside the hogan, walking clockwise around the altar. The singing continued through the small hours of the morning. The peyote, now only in the bottom of the nearly empty pot, was passed a time or two more, and those who wished to have more took whatever they wanted. However, few took any after midnight, meaning that most participants ate five to seven pieces during the night-long meeting.

The hours began to weigh heavily on me, and I could see that a few others had also become drowsy. The young children had gone to sleep, but most active participants sat contentedly with a somewhat relaxed, transfixed look upon their faces. A few smiled, some frowned, and others had expressions that continually changed. No one was "intoxicated," unconscious, or violent. Most people seemed introspective and withdrawn into their own worlds of thought; however, when their turn came to sing or drum, they demonstrated their awareness of what was going on in the room by immediately participating, both appropriately and with self-control. In fact, some drummers would even retune the water drum by sloshing water onto the inside of the buckskin or blowing under the corner of it to get a better sound.

About 3:00 A.M. the Kiowa Cedarman announced that there was going to be a curing ceremony. He said that one of the young women present was lame, and that medicines given to her at the hospital had not helped; therefore, he, a Kiowa medicine man, would cure her by sucking out the evil spirits. The tall Kiowa arose, went to the altar, picked up a hot coal in his fingers, and placed it between his teeth. He blew air out of his lungs, and the coal brightly glowed in the dim light of the hogan. Suddenly he popped the hot coal into his mouth and closed it. The patient was lying on her stomach near the door with a friend sitting beside her holding her hand. The Kiowa got down on his knees beside the girl, carefully pulled up her blouse to uncover the small of her back, placed his mouth on part of the exposed skin, and began to suck strongly. Most of the people present seemed somewhat impassive about this amazing effort to suck the evil spirit out of the patient. Probably many had seen similar actions by other medicine men. When the coal became cold, the Kiowa spat it into his hand and placed it back on the altar. He instructed the young woman to turn over and sit up; he then repeated the procedure of placing a hot coal in his mouth, but this time sucked strongly and noisily on her forehead to draw out more of the evil spirit. From time to time he rinsed out his mouth with water from a small container and then spat into a small can. Apparently he wanted to wash the evil spirits out of his mouth. While the Kiowa was performing the cure, some of the members sang and offered a prayer in the patient's behalf. Finally, he instructed the young woman to return to her place, and he walked back to his position to the left of the Roadman.

The singing resumed, and somewhat later we had another chance to go outside to stretch. Soon after we returned, I noticed the moon through the smoke hole in the roof of the hogan. About 4:30 A.M. the day began to dawn, and the sky, visible through the smoke hole, became lighter and lighter. About 5:00 A.M. a woman participant arose and left the hogan alone. While she was gone, another member took a Pendleton blanket from a hanger on the wall and placed it on the floor east of the altar. Soon the woman returned carrying the enameled bucket of water; she placed it between the altar and the blanket, and knelt on the blanket. After being handed cigarette makings, she rolled a cigarette, lit it with the smoke stick, and began to slowly and deliberately puff on it. As she smoked, she began to pray in Navajo, from time to time stopping to relight her cigarette during the long, increasingly emotional prayer. Tears soon began to flow down her cheeks, and again other members began to sob and grunt. Occasionally, however, I heard a few giggles and saw some of the men grinning. Afterwards I learned she had become very specific in confessing some of her past sins and had made several strong remarks favoring women's liberation, much to the amusement of some of the men. This moment of levity quickly passed back into one of great seriousness when she asked the Almighty for forgiveness of her sins. When she had finished praying, the water bucket was again passed as at midnight, and by the time all had drunk, the sun had risen and bright light streamed in through the partially opened door.

The Kiowa Cedarman announced to us that it was time for the closing ceremony, and the Roadman began to sing the four Closing Songs. The Kiowa then stood and began to pray again in English, becoming more and more emotional and eliciting many favorable grunts and sobs from the group. At the end of the prayer the Fire Chief began to change the shape of the coals and ashes for the final part of the meeting by putting a forked tail on the Peyote Bird. Four containers were then brought into the hogan: the bucket of water, a dish of canned corn, a dish of ground meat, and a dish of raisins and canned fruit. These containers, along with a collection cup, were placed in a row from the altar towards the door. The Roadman offered prayers for the food, and the containers were passed clockwise around the room with participants serving themselves with their fingers or spoons. The collection cup followed the food dishes around the room, and most participants put in some coins or bills to help pay for the meeting.

When all had eaten of the peyote breakfast, the Roadman, Chief Drummer, Cedarman, and Fire Chief began to put away the paraphernalia. The Chief Drummer took apart the drum and drank the water inside, while the Fire Chief picked up the cigarette butts, carried them outside, and threw them in the fire. After all the ritual apparatus had been put in the special box, the Roadman announced that anyone could now speak if he or she wished. The sunlight streamed into the hogan through the open door, and the people began to smile, whisper with one another, and express publicly their appreciation for the meeting. There seemed to be a greater closeness among those present than earlier that night. Many thanked the Roadman for leading such a good meeting and complimented the Kiowa guest on his curing ceremony and instructions. Others thanked the host for sponsoring the meeting, and he expressed his deep appreciation for their prayers on his behalf. A few others took this opportunity to make public confession and to promise to lead better lives with no drinking or immorality. Finally, the Roadman declared that the sponsor would be successful in his studies because all had gone well in the ceremony. He stood up and walked out of the hogan with the rest of us following him.

The peyote meeting was over, and outside we warmly shook hands with our fellow worshipers; we were now friends because we had smoked cigarettes and eaten peyote together. As we stood talking, the Fire Chief came out of the hogan carrying a metal tub containing the sand and ashes from the altar; he spread them near the door of the hogan.

About an hour later all of the participants and their families were invited to reenter the hogan to share in a feast provided by the sponsor. Again we sat on the mattresses, but now the center of the hogan was covered with table cloths and an interesting variety of food: cold mutton, mutton ribs, soup, Navajo fry bread, rolls, potato salad, green salad, coffee, canned fruit, cookies, and crackers. We began the feast by drinking from a bucket of water passed around the room, only this time we used individual paper cups rather than a single enameled cup. We then ate heartily from paper plates with plastic forks and spoons. When all had eaten, individuals and families began to depart; each person slowly walked clockwise around the room and shook hands with everyone present. When it came our turn to leave, we stood up, picked up our hats, smiled, and warmly shook hands with our new Native American friends with whom we had shared smoking, praying, singing—and eating peyote.

Comments about the Navajo V-Way Ceremony

The Navajo ceremony just described had elements from the Christian religion besides the pictures on the wall of Jesus and the Last Supper. Peyote was spoken of as "the sacrament," and the staff had a cross near its tip. Continual references were made to sin, confession, forgiveness, and Almighty God. However, the name of Jesus was rarely mentioned. This Navajo ceremony, like those of the Plains, is a combination of traditional Native American beliefs and Christianity, the purely indigenous elements being the musical instruments, references to witches and spirits, and the curing ceremony.

The methods of financing a Navajo peyote meeting are quite simple from what we observed and were told. The Kiowa emphasized that neither the Roadman nor any of the other officers should ever charge a fee for conducting a ceremony. He claimed that they should willingly participate and receive only what the sponsor and participants want or can afford to give to them. Some Roadmen, he said, led meetings only to make money, but this is wrong.

The sponsor of the meeting is expected to pay for the food and any other items involved with the ceremony. The sponsor of our meeting paid about $100 for the Roadman and the food, and the money from the collection cup (about $27) was given to the other officers who participated in the service. Many people commented that our sponsor had provided an especially good feast.

One of the most remarkable aspects of the ceremony was the emotional involvement of the participants. The prayers deeply affected those present, as evidenced by the sobs, grunting, and weeping. Moreover, many participants felt a genuine need for repentance. After the ceremony, one man swore that he was going to quit drinking and would more sincerely follow "the Peyote Way." He commented that when he ate peyote, it made him so sick that he had to leave the hogan. "All the evil and bad alcohol came out of my body," he declared. "I have been cleansed by peyote and now want to live a better life." This man, like many others, had gone through what was to him a deep and very meaningful emotional experience. He was willing to confess openly his sins of the past and to declare that he would try to be a better person hereafter.

Visions did not seem to play an important part in the ceremony. Few participants admitted that they had ever had visions, and judging from the amount of peyote consumed during the night, it is unlikely that many could have had them. One young, well-educated woman confided that she had had no visions, nor did colors appear to be altered in any special way. However, she felt that the peyote had given her a greater perception of people by somehow removing the facade or mask of impersonality. She smiled and said that for the first time she was able to see into the true nature or character of the people participating in the ceremony.

The peyote ceremony was also noteworthy for the way it fostered the spirit of comradery. Most of the major parts of the ceremony—incense blessing, smoking, consuming peyote, singing, praying, drinking, and eating—made one feel a part of the whole. At no time after the ceremony began did I, an Anglo, feel uncomfortable or unwelcome, and several times prayers were offered for greater understanding, friendship, and love between "red and white men."

For many Native Americans, the peyote ceremony has considerable influence, offering "prayer for his soul, food for his stomach, health for his body, prestige, and self-expression. It allows for the venting of certain aggressive feelings and, to a degree, soothes the ever-present sense of cultural loss" (Arth 1956, 28). The ceremony is part of a serious, valid, and meaningful religion — a demanding, moralistic faith that requires much emotional and physical involvement. The ceremony itself, we were told, is a means or a tool by which believers can find and attune to God through the sacramental use of peyote.

CHAPTER FOUR

The Peyote User's Experience

Humans have used peyote because of the way it makes them feel, think, and act. Although it may produce maladaptive responses such as psychotic or panic reactions, escape from responsibilities and inner tensions, or depersonalization, on the other hand it may result in efforts to acquire new experience or knowledge, and to reduce one's inner conflicts in order to function more constructively and safely in society. To some it is as if this "divine cactus" were actually transporting part of them to another world. In the ceremonies of the Native American Church, peyote is the prime element or sacrament whereby the participants can communicate with God.

This chapter discusses what people have experienced when they have eaten peyote or have taken its alkaloid mescaline, but in contrast to the previous chapter's emphasis on the peyote experience of Native Americans in a religious context, this chapter deals mainly with the experiences of Europeans and European immigrants.

Aldous Huxley was one of the most notable and literate admiring users of peyote. When describing his experience with mescaline and the other world to which it transported him, Huxley (1956) wrote that "the classic mescaline experience is not of consciously or unconsciously remembered events, does not concern itself with early traumas, and is not, in most cases, tinged by anxiety and fear. It is as though those who were going through it had been transported by mescaline to some remote, non-personal region of the mind" (46). One may actually feel an almost mystical loss of oneself into something larger and more significant than would be possible under normal, or non-peyote, circumstances.

There is such a wide variety in accounts of the peyote or mescaline experience, and apparently so few common elements, it is difficult to analyze. However, the experience can be examined fairly well by dividing it into separate phases. One of the major problems in understanding—and describing—the peyote or mescaline experience is the subject's difficulty in communicating what he or she has actually gone through. There seem to be two main reasons for this: (1) it is usually hard to communicate a deep esoteric experience, and (2) peyote and mescaline cause a disorientation of the senses that results in the loss of the usual landmarks by which we communicate; for example, space relationships and the perception of time are greatly distorted.

Peyote produces a mental condition that psychiatrist Arnold M. Ludwig (1969) describes as an "altered state of consciousness" (10–13). Such mental states may be induced by a variety of drugs, as well as by certain environmental experiences. Mental aberrations similar to those induced by peyote may be caused by such influences as solitary confinement, brainwashing, ecstatic trances, intense praying, hypnosis, autohypnotic trances, and dehydration. The ingestion of a variety of other pharmacological substances—stimulative, psychotomimetic, and sedative in nature—can also produce an altered state of consciousness similar to that caused by peyote. At first view it would seem unlikely that all of these events and drugs could result in states of consciousness having similar characteristics. However, Ludwig (1969, 13–17) and others suggest that most altered states of consciousness are characterized by certain recognizable features: alterations in the thinking process, a disturbed time sense, loss of self-control, change in emotional expression, distortions in body image, distortions in perception, increased meaning or significance to experiences, an inability to communicate, feelings of new hope or rejuvenation, and hypersuggestibility. Most of these features are found in the peyote experience.

The altered state of consciousness described by Ludwig could also be called a delirium, a dysfunction of the brain that is only temporary and can be reversed. Delirium, which has also been called "acute brain syndrome," is usually classified by its probable causes: such things as disorders of metabolism, trauma, systemic infection, disturbance of circulation, and substance intoxication and withdrawal (American Psychiatric Association 1980, 104–7). It is the deliria caused by intoxication that are of the most interest, because these would include the action of psychotomimetics such as peyote. The symptomatic expressions of this particular brain syndrome vary to a considerable degree, especially with respect to the individual's state of consciousness (Kolb 1977, 681–82). Peyote, for example, does not cloud the consciousness.

Some of the probable causes of the altered state of consciousness from psychotomimetics such as peyote will be considered in chapter 6, but it is important to emphasize here that no drug experience can be wholly attributed to a pure chemical effect. The effects of peyote (or mescaline) may be modified by any or all of the following factors: (1) the form in which peyote or mescaline is taken and the dosage, (2) the personality and present mood of the subject, (3) the environmental setting or place where one has the experience, and (4) the preparation or prior conditioning of the subject. On occasion, the personality can be overridden by either explicit or implicit suggestions. The Roadman of the Native American peyote ceremony, as well as the tone and nature of the singing, may "direct" particpants' sensory experiences. The same is true of people who take peyote in a nonritualistic setting; companions may direct their sensory perceptions either purposely or indirectly. The setting also may have considerable power of suggestion. A sterile hospital room, for example, may evoke quite different responses from an individual than a darkened room with a flickering fire and rhythmical music.

Another problem in analyzing the peyote experience is that the effects of ingesting the whole plant with its many alkaloids and taking pure mescaline seem to differ (Schultes and Hofmann 1980, 200). There have been relatively few carefully reported experiences of taking peyote, and frequently Native Americans do not wish to describe their experiences. Thus, most of the descriptions to follow are those of persons who took pure mescaline. Both peyote and mescaline produce brilliantly colored visions and many other sensory effects, but the impact of the whole peyote plant tends to be more complex, variable, and unpredictable.

Heinrich Klüver (1966, 51), a behavioral biologist from the University of Chicago, noted that Kurt Beringer and other early European researchers on peyote felt that there were three main constituents of "the mescal psychosis": (1) abnormal sensory phenomena, (2) a fundamental alteration in the conscious states and attitudes, and (3) abnormal emotional states. These three constituents demonstrate a disorganized mental state characteristic of altered states of consciousness induced by drugs and toxins. The integration of intellectual thought is deeply disturbed, and there is a disturbance of association, a blocking of voluntary intellectual thought, and extreme distractibility. People who have taken substances such as peyote or mescaline have commented that it is difficult to concentrate because there is such a rapid flow of ideas and sensations; moreover, there is a blurring of the distinction between cause and effect, an impairment in abstract thinking, and difficulty in the power of expression.

Motor activities are also affected by peyote. Often the subject delights in laziness and is overcome by lethargy and a strong desire not to engage in any active movement. Probably some of this tendency to remain immobile is because people simply have problems making up their minds what to do. Some individuals report impulsive outbursts of meaningless activity alternating with the lethargic periods.

Another characteristic of the "toxic delirium" also experienced by peyote users is a state of euphoria that may alternate with anxiety, excitement, or bewilderment. Whether high spirits—or fear and anxiety—predominate may be influenced by the subject's personality (Naditch 1974, 395). Usually, however, most peyote experiences produce a state of mental exhilaration. This interesting euphoria-anxiety relationship is one of the most impressive parts of the peyote experience, perhaps because it can be so frightening. The individual's ego actually seems threatened by the altered perception of his or her own body and the surrounding environment. During this phase subjects may display either fear or anger depending on how they are able to cope with the situation, and these conflicting feelings of euphoria and anxiety may produce a set of experiences that is a joy—or a terror. Raymond Mortimer reported that

> the inability to feel suddenly became alarming in the extreme. Had the drug merely paralyzed my emotions, I wondered, or had it perhaps destroyed them forever? Life without them would be unendurable. Then I decided that this terror was itself an emotion, and that its appearance meant that I was emerging from the influence of the drug. This reasoning brought momentary relief, but I now began to suffer a generalized apprehension amounting to panic. I was not frightened of pain or death; I was frightened only of continuing to exist in my present unmotivated anguish. Though my brain was lucid, I was suffering the torments caused by some forms of madness. (Quoted in Zaehner 1961, 211)

Phases of the Experience

Peyote, or mescaline, profoundly affects people who take it because it creates within them a form of "toxic delirium" manifesting most of the features of the altered state of consciousness Ludwig described. Not only are the senses altered, but there may even be a depersonalization or dissociation of the intellectual part of the personality from the rest of the mind. Yet even though the peyote experience is said to be "paranormal" in its effects—beyond or outside the range of normal personal experience (Unger 1963, 114)—one of the remarkable characteristics of peyote is that there is little clouding of the consciousness.

The peyote experience can be divided into two stages, in which the "hangover precedes the ebriety" (Jacobsen 1963, 485). In other words, first there is a phase dominated by bodily symptoms, followed by one of psychic or mental manifestations (Dixon 1899/1900, 79–81).

Phase 1

Although this stage may be strongly influenced by many of the alkaloids present in peyote, mescaline seems to be the most important. After a latency period of about an hour (after oral ingestion), one begins to experience a series of disagreeable symptoms such as nausea, vomiting, dizziness, sweating, palpitations, and headache. Additional effects may sometimes include feelings of heat, cold, and choking; chest and neck pains; shortness of breath; a "peppermint" taste in the mouth; hunger; stomach cramps; pupil dilation; tremors; an urgency to urinate; and generalized restlessness and discomfort. During this stage some people fear that they are dying and therefore express intense anxiety and fear. Occasionally this anxiety is shown as severe agitation and even violence. This first phase may persist for up to three or four hours. Usually the more adverse symptoms subside prior to the onset of phase 2. Phase 1 can be most accurately described as one of depression, anxiety, and physical discomfort.

Phase 2

Usually the second phase quickly follows the first and is characterized by euphoria and elation. This psychic or mental phase may produce delightfully happy and dreamy feelings, as well as pleasant fantasies. Subjects may even experience strange delusions of grandeur and think they possess increased physical power and ability. As this phase is entered, a person may have the paradoxical experience of feeling both depressed and elated at the same moment.

Phase 2 may also be divided into two stages. After a preliminary stage of exhilaration, there is a second stage of "intoxication" characterized by profound psychic experiences including visions, a distortion of sense perceptions, and a contemplative mood. All senses may become distorted, or perhaps only one or two. Subjects may also experience "synesthesia" of the senses, with the stimulation of one sense resulting in sensations of others.

Visions or hallucinations are the most impressive element of the peyote experience, but not everyone sees them. These and other sense distortions tend to be entertaining and may even be humorous. People also report that thoughts and images seem to pour out so fast that before one is fully developed, another is there. Most people have good recall of the experience, although details are often blurred or omitted. Few experience any sense of fatigue even though the total peyote experience may last for eight to ten hours.

Somatic phenomena that occur during this second phase are incoordination, twitching of muscles, increased body reflexes, occasional tremors, blunting of painful and tactile sensations, and widely dilated pupils that tend to react sluggishly to light.

Important Aspects of the Experience

The literature that describes the effects of peyote frequently mentions certain important features such as the feeling of a dual existence ("depersonalization"), distortion of time and space, difficulty in communicating, effects on memory and thinking, a mixing and heightening of the senses, emotional extremes, inhibition of sex drive, and an absence of dreams. A comparison of these aspects with the features Ludwig (1969, 13–18) found to characterize the altered state of consciousness demonstrates that the peyote experience is an excellent example of the near-psychotic state that may be temporarily induced by drugs and toxins. By far the most noteworthy effect of peyote is on the visual sense; this will be dealt with in the next major section of this chapter. What follows here is a brief consideration of other important aspects of the peyote experience, many of them illustrated by personal accounts.

The Feeling of a Dual Existence

One of the strangest effects of peyote and mescaline has to do with one's perception of oneself. There is a strong subjective experience in which people find that they are no longer themselves. They feel as if something strange has been substituted for their personality; that is, they have the distinct impression that bodily unity has been lost and that they now have a dual existence. William Braden (1967), a reporter for the *Chicago Sun-Times* and a contributor to *Saturday Review* and other contemporary periodicals, recalled that while he was under the influence of mescaline, "I looked down at my tweed jacket and my green-and-white striped shirt, then farther down at my crossed legs. Black pants, black socks, black shoes. The only trouble was, the legs weren't mine. They looked alien and somehow sinister. I knew they were attached to my body and I knew I could move them, but they weren't my legs. They weren't me. Or better yet, I wasn't *them*" (232). The subject suddenly realizes that his or her perceptions do not exist within the world of physical reality. E. Robert Sinnett (1970), a psychologist who experimented with mescaline, stated: "My self seemed like a small sphere within the center of my head. The rest of my body seemed somehow peripheral and empty, as if my self were a pea-sized object in a shadowy, gray void enclosed by my body" (31).

This strange subjective feeling of unreality may be unpleasant and upsetting because the subject encounters so little in the way of familiar landmarks. It is usually frightening to have the feeling that you have been transposed into someone—or something—else! A subject of the European researchers Erich Guttmann and W. S. Maclay (1936) commented that as he drew his hand across his face, "it felt as if the hand had no connexion with me and did not belong to me. I looked at it as if it was not my own, then the feeling faded and not even by using my imagination could I recapture it" (194).

For some people, depersonalization makes them feel like machines; actions seem to be mechanical and there is an absence of willpower. Professor R. C. Zaehner (1961) of Oxford University said, "I had no difficulty at all in walking, but had a curious sensation that my body was under perfect control, which seemed odd, for I was certainly, I thought, in no position to control it. The body seemed momentarily to be leading an autonomous life of its own,—and very well it managed it,—whereas 'I' was becoming increasingly confused and unsure of myself" (216).

The effect may even seem humorous and magical—but at all times overwhelming because of the lack of "normal reality." Another subject of Guttmann and Maclay (1936) remarked that "as I hold this pencil, which seems suspended . . . and feels like a piece of india-rubber, I feel just like Alice in Wonderland. This room and myself feel one. When I drank tea it seemed silly to pour it down through the centre of the room. I feel miles away from the paper—and why should I bother" (209).

Distortion of Time and Space

Subjects under the influence of peyote or mescaline relate that their experiences seem to take place outside of time. Especially with heavy doses of the whole plant or of mescaline, strange illusions of time occur, and an effect is created in which a new dimension seems to be added. What may seem like hours to the subject may be only a few minutes on a clock. One of Dr. Guttmann's subjects said that "time seemed to stand still. I had a feeling of relief when I heard that it was five minutes later. I was dreadfully impatient. I was the more embarrassed that it was impossible for me to follow the small hand of the watch or to count my pulse to orientate myself." Another commented, "I drank a spoonful of soup, looked around me, and looked down again at my plate; it had been in front of me for hundreds of years" (quoted in Guttmann 1936, 209).

Space perception is often so distorted that one's experiences seem to lack reference to physical reality; space may seem to become so extended that it is infinite. Humphrey Osmond (1970), a psychiatrist and active researcher on psychotomimetic drugs, remarked in his article titled "On Being Mad" that "at one moment I would be a giant in a tiny cupboard, and the next, a dwarf in a huge hall. It is difficult enough to explain what it feels like to have been Gulliver, or Alice in Wonderland, in the space of a few minutes, but it is nearly impossible to communicate an experience which amounts to having been uncertain whether one was in Brobdingnag or Lilliput" (26–27).

Aldous Huxley experienced both time and space distortion:

And along with indifference to space there went an even completer indifference to time.

"There seems to be plenty of it," was all I would answer when the investigator asked me to say what I felt about time.

Plenty of it, but exactly how much was entirely irrelevant. I could of course, have looked at my watch; but my watch, I knew, was in another universe. My actual experience had been, was still, of an indefinite duration or alternatively of a perpetual present made up of one continually changing apocalypse. (Huxley 1959, 20)

Samuel W. Fernberger experienced similarly profound effects of time and space distortion more than seventy years ago:

> Accompanying this distortion of space was also a distortion of time. A given period of time seemed very much more extended than normally. Speech seemed slow. Walking became a ponderous affair. In walking I felt successively sensations from the muscles and joints involved. I became aware at each step, for example, of sensations due to the curling of the toes. If the clearness of the sensations accounts for the distortion of space, the very rapidly changing focus of attention would readily account for the distortion of time. We naturally and normally would judge a period of time by the "filling," by the number of processes that have claimed the center of attention. Under the influence of peyote this filling, for a given unit of time, is much increased, and so the period seems extended. (Fernberger 1923, 269)

Difficulty in Communicating

Most people under the influence of peyote or mescaline have a reduced capacity to communicate. Observers feel particularly frustrated when trying to learn what a person is experiencing because subjects may either say nothing, or what they try to talk about may seem almost indescribable, despite the fact that the images and thoughts are vivid and visual. The difficulties of describing the experience are caused by the absence of appropriate associations, the flooding of the mind or senses with too many things at once, the inability to say what one wants to say, excessive preoccupation with one's experiences so that there is simply no desire to communicate with others at the moment, the mistaken idea that communication has already taken place when in fact it has not, and problems inherent in talking about a deeply moving spiritual or mystical experience. It is not surprising that the combined influence of all these factors, each one of which is enough by itself to cause difficulties in communication, makes it extremely difficult for peyote users to relate the experience to others in any meaningful way.

Effects on Memory and Thinking

During the peyote experience, the subject appears to possess a good memory, but afterward there tend to be gaps in what can be recalled. Probably this limited recall is caused by the inability of the subject's mind to filter out, focus on, and slow down the many images and thoughts as they crowd through the consciousness. Apparently the mind becomes supersaturated, and the brain less able to organize and store memories for later recall. Nonetheless, the impact is still there, and the subject usually wants to share the experience when the effects of peyote have worn off.

Mixing and Heightening of the Senses

A mixing of the senses, called synesthesia, is the excitation of one sense organ by a stimulus that produces a train of images in one or more other sense organs. It is one of the most remarkable effects of peyote and mescaline, almost as if a "sixth sense" were created, for one experiences such things as hearing color or seeing sound (Marks 1975, 317). There is little that is equivalent to this type of sensation in our normal experiences. Subjects have commented that "whenever I touch something, I have light sensations" or "the barking of a dog moved the whole picture and vibrated through my right foot" (quoted in Klüver 1966, 50). Although the most common type of synesthesia is visual-auditory, there may also be visual-gustatory (with tastes), visual-tactile, visual-olfactory (with smell), visual-kinesthetic (muscular), and visual-thermal. Some people even experience visual-algesic synesthesia, or visual imagery with pain. A subject of Lorna Simpson and Peter McKellar (1955) of Aberdeen University in Scotland, when pricked on her hand, remarked, "I've got concentric circles like round the top of a radio mast. As you touch me, jagged things shoot up, little sort of jagged things shooting out from a center" (146).

As well as being the most common, some of the most spectacular impressions of synesthesia are in the realm of the visual-auditory. William Braden wrote that when he put on the earphones of a phonograph,

> A majestic Beethoven chord exploded inside my brain, and I instantly disappeared. My body no longer existed, and neither did the world. The world and I had been utterly annihilated. I could feel the pressure of the earphones; but in the space between the phones, where my head should have been, there was absolutely nothing . . . *nothing!* I was Mind

alone, lost in an icy blue grotto of sound. There was only the music, and then bright colors that turned out to be musical notes. The notes danced along a silver staff of music that stretched from one eternity to another, beyond the planets and stars and space itself. Red notes. Blue notes. But they had no substance or dimension, and nothing was real in that empty cavern between the two earphones. That unbounded abyss. The music rolled on in orgiastic waves of sound and color, and then I myself was one of the notes. I was being swept along on the silver staff, at twice the speed of light, rushing farther and farther away from my home back there in the Milky Way. In desperation, at the last possible moment, I reached up with hands I did not own, and I tore off the earphones. (Braden 1967, 236)

As mentioned earlier, in contrast to many other drugs such as alcohol and narcotics, peyote and mescaline do not greatly disturb the general thought processes or cloud the consciousness unless there has been a heavy dose. Thought disturbances that do occasionally occur can be characterized as a slowness and difficulty of concentration, a flight of ideas, limited language disorders, and a slowing of speech. On balance, however, it appears that peyote tends to enhance some aspects of mental thought and perception, so that, as Vincenzo Petrullo (1934) observed, "colors appear brighter, surfaces show greater details, noises sound louder, and in the case of drums beating considerably richer in tone; speech has greater meaning, logic is more acute, ideas are of a clearer essence; in short, the mind is alert and more sensitive to stimulus from the objective and subjective worlds" (7).

Emotional Extremes

Many people who take peyote or mescaline have profoundly emotional experiences. Minor events or objects take on great significance, and in some cases individuals believe they have had truly mystical experiences. Aldous Huxley, in his marvelous little book *The Doors of Perception* (1959), was swallowed up in a mystical experience when he observed a simple vase of flowers.

The vase contained only three flowers—a fullblown Belle of Portugal rose, shell pink with a hint at every petal's base of a hotter, flamier hue; a large magenta and cream-coloured carnation; and, pale purple at the end of its broken stalk, the bold heraldic blossom of an iris. Fortuitous and provisional, the little nosegay broke all the rules of traditional good

> taste. At breakfast that morning I had been struck by the lively dissonance of its colours. But that was no longer the point. I was not looking now at an unusual flower arrangement. I was seeing what Adam had seen on the morning of his creation—the miracle, moment by moment, of naked existence. (Huxley 1959, 16–17)

Even the books and furniture of his room produced feelings and ideas of profound significance:

> The books, for example, with which my study walls were lined. Like the flowers, they glowed, when I looked at them, with brighter colours, a profounder significance. Red books, like rubies; emerald books; books bound in white jade, books of agate, of aquamarine, of yellow topaz; lapis lazuli books whose colour was so intense, so intrinsically meaningful, that they seemed to be on the point of leaving the shelves to thrust themselves more insistently on my attention. . . . I was looking at my furniture, not as the utilitarian who has to sit on chairs, to write at desks and tables, and not as the camera-man or scientific recorder, but as the pure aesthete whose concern is only with forms and their relationships within the field of vision or the picture space. But as I looked, this purely aesthetic Cubist's-eye view gave place to what I can only describe as the sacramental vision of reality. I was back where I had been when I was looking at the flowers—back in a world where everything shone with the Inner Light, and was infinite in its significance. The legs, for example, of that chair—how miraculous their tubularity, how supernatural their polished smoothness! I spent several minutes—or was it several centuries?—not merely gazing at those bamboo legs, but actually being them—or rather being myself in them; or to be still more accurate (for "I" was not involved in the case, nor in a certain sense were "they") being my Not-self in the Not-self which was the chair. (18–21)

Perception of extraordinary experiences such as Huxley's vividly described "color-symphonies" affect some people to such a degree that peyote has become a method of communication, a means of revelation, with God. Although not all societies consider visions and other emotional revelations desirable, many indigenous people throughout the world have deliberately sought them as a means by which they may come in contact with the divine (Bourguignon 1977, 10–21).

On the other hand, peyote or mescaline may produce an emotional response quite opposite to the mystical—even as a reaction to quite ordinary visual stimuli. When R. C. Zaehner looked at an ordinary electric lightbulb,

> it seemed to grow brighter and to expand a little. On shutting my eyes the after-image behaved on more or less conventional lines,—starting as green in the center of my visual field, it ascended, appeared to explode, became red in the middle and green outside. It changed colour so often, exploded into a dim bluish pattern and reformed again so often that I could not describe these metamorphoses quickly enough. The image that remains most clearly in my memory is that of a slowly mounting fiery ball which reminded me of an atomic explosion. I thought it sinister and described it as "horrid." (Zaehner 1961, 213)

Inhibition of Sex Drive

Persons under the influence of peyote or mescaline seldom have an interest in sex. It can probably be said that peyote is neither an aphrodisiac or anaphrodisiac simply because the subject under the influence of peyote is too involved with other things to be concerned with sexual matters. However, since peyote only changes the perception of events in a way that depends on the setting and the personality of the user, it is quite possible for some people to have erotic experiences. Some writers, in fact, report events in which peyote provided a strong urge for a sexual experience. These desires seem to be the exception rather than the rule, however, because the locale and mind-set of people under the influence of peyote are not usually oriented in the direction of sex.

Absence of Dreams

One might expect peyote to cause vivid dreams since it produces such strong visions and other sensory illusions in people who are awake, but this simply is not the case. Several individuals who have experienced peyote commented that dreams are rare during sleep following the peyote experience; in fact, it seems that peyote may in some way inhibit or replace dreams for a period of several days.

VISIONS

The visual experience of peyote is the climax of all the effects the cactus has upon humans and their senses. Peyote permits one to invade the world of the unknown in a visual encounter of mind-expanding—or mind-shattering—dimensions.

Some of those who have commented on the visual effects of peyote, or of the alkaloid mescaline, refer to them as hallucinations. Others disagree, stating that most who experience the effects of peyote can distinguish them from physical reality (Barron, Jarvik, and Bunnell 1964, 29). A hallucination is perception without the appropriate stimulation of either internal or external sensory receptors. In other words, one experiences something that is not really present, but is unable to distinguish these perceptions from what is physically real. The question, then, is whether those who experience the effects of peyote or mescaline can readily distinguish the visual experience as unreal. The term illusion—a false interpretation of external reality—has been suggested as descriptive of some aspects of the peyote experience (Bliss and Clark 1962, 92). Others have proposed the term "pseudo-hallucination" (Malitz, Wilkens, and Esecover 1962, 52). Schultes and Hofmann (1979, 12–13) extensively discuss the many terms that have been coined to describe the effects of peyote, mescaline, and similar drugs. They prefer to use the term "psychotomimetic," which I will follow. Despite this great variety of terms, few researchers question the remarkable sequence of events and the general effects of peyote in the production of what I shall simply call "the visual experience."

Most people who take peyote or mescaline have simple visions containing only abstract forms, lights, colors, and flashes. Some of them also have more-complex visual experiences in which familiar objects such as landscapes, animals, or human beings are seen. These images rapidly and continually change but can be recognized and described.

Many people have reported similar visual experiences with peyote or mescaline. One of the classic descriptions of the peyote experience was published in 1897. In the following portion of his account, the great English psychologist and writer Havelock Ellis mentions many of the general visual effects, which will subsequently be examined in more detail.

> On Good Friday, being entirely alone in quiet London rooms, I made an infusion of three buttons (a full dose) and took it in three portions

at intervals of an hour between 2:30 and 4:30 P.M. The first noteworthy result (and the only one of therapeutic interest which I have to record) was that a headache which had been present for some hours and showed a tendency to aggravation was immediately relieved and speedily dissipated. There was slight drowsiness before the third dose was taken, but this speedily passed off and gave place to a certain consciousness of unusual energy and intellectual power, which also quickly passed off, and was not marked and prolonged, as with Dr. Weir Mitchell. So far no visual phenomena had appeared, even when the eyes were closed for several minutes, and there was yet no marked increase of knee-jerk; there was, however, a certain heightening of muscular irritability, such as may be noted when one has been without sleep for an unusual period. The pulse also began to fall. After the third dose I was still feeling on the whole better than before I began the experiment. But at 5 P.M. I felt slightly faint, and it became difficult to concentrate my attention in reading; I lay down and found that the pulse had now fallen to 48, but no visual phenomena had yet appeared. At 6 P.M. I noticed while lying down (in which position I was able to read) that a pale violet shadow floated over the page. I had already noted that objects which were not in the direct line of vision showed a tendency to be heightened in colour and to appear enlarged and obtrusive, while after-images began to be marked and persistent. At 6 P.M. there was a slight feeling of faintness as well as of nausea, and the first symptoms of muscular incoordination began to appear, but there was no marked discomfort. By 7 P.M. visions had begun to appear with closed eyelids, a vague confused mass of kaleidoscopic character. The visual phenomena seen with open eyes now also became more marked, and in addition to the very distinct violet shadows there were faint green shadows. Perhaps the most pleasant moment in the experience occurred at 7:30 P.M., when for the first time the colour visions with closed eyes became vivid and distinct, while at the same time I had an olfactory hallucination, the air seeming filled with vague perfume. Meanwhile, the pulse had been rising, and by 8:30 had reached its normal level (72 in the sitting posture). At the same time muscular incoordination had so far advanced that it was almost impossible to manipulate a pen, and I had to write with a pencil; this also I could soon only use for a few minutes at a time, and as I wrote a golden tone now lay over the paper, and the pencil seemed to write in gold, while my hand, seen in indirect vision as I wrote, looked bronzed, scaled, and flushed with red. Except for slight nausea I continued to feel well, and there was no loss of mental coolness or alertness. When gazing at the visions with closed eyes I occasionally experienced slight right frontal headache, but as I only noticed it at these times I attribute this mainly to the concentration of visual attention. In one very important particular my experience differs from Dr. Weir Mitchell's. He was unable to see the visions with open eyes even in the darkest room. I found it perfectly easy to see them with open eyes in a dark room, though they were less brilliant than when the eyes were closed. At 10 P.M., finding that movement distinctly aggravated the nausea and faintness, I went to bed, and as I undressed was impressed by the bronzed and pigmented appearance of my limbs. In bed the nausea entirely disappeared, not to reappear, the only discomfort that remained being the sensation of thoracic oppression, and the occasional involuntary sighing, evidently due to shallow respiration, which had appeared about the same time as the vision began. But there was not the slightest drowsiness. This insomnia seemed to be connected less with the constantly shifting visions, which were always beautiful and agreeable, than with the vague alarm caused by thoracic oppression, and more especially with the auditory hyperaesthesia. I was uncomfortably receptive to sounds of every kind, and whenever I seemed to be nearly falling asleep I was invariably startled either by the exaggerated reverberation of some distant street noise (though the neighbourhood was even quieter than usual), or, again, by the mental image (not hallucination) of a loud sound, or, again, as I was sometimes inclined to think, by actual faint hallucinatory sounds; this, however, was difficult to verify. At a later stage there was some ringing in the ear. There was slight twitching of the larger muscles of the legs, etc., and before going to bed I had ascertained that the knee-jerk was much exaggerated. The skin was hot and dry. The visions continued. After some hours, tired of watching them, I lighted the gas. Then I found myself in a position to watch a new series of vivid phenomena to which the previous investigators had not alluded. The gas—i.e., an ordinary flickering burner—seemed to burn with great brilliance, sending out waves of light which extended and contracted rhythmically in an enormously exaggerated manner. What chiefly impressed me, however, were the shadows which came in all directions, heightened by flushes of red, green, and especially violet. The whole room then became vivid and beautiful, and the tone and texture of the whitewashed but not remarkably white ceiling was immensely improved. The difference between the room as I then saw it and its usual appearance was precisely the difference one may often observe between the picture of a room and the actual room. The shadows I saw were the shadows which the artist puts in, but which are not visible under normal conditions of casual inspection. The violet shadows especially reminded me of Monet's paintings, and as I gazed at them it occurred to me that mescal [peyote] doubtless reproduces the same conditions of visual hyperaesthesia, or rather exhaustion, which is certainly produced in the artist by prolonged visual attention (although this point has yet received no attention from psychologists). It seems probable that these predominantly violet shadows are to some extent conditioned by the dilatation of the pupils, which, as the American observers had already noted, always occurs in mescal intoxication. I may remark in this connexion that violet vision has been noted after eye-operations; and Dobrowolsky has argued that a necessary condition for such vision is the dilatation of the pupils produced by atropine, so that the colour vision (chiefly violet, though to some extent of other colours) is really of the nature of an after-image due to bright light. Dobrowolsky's explanation seems to fit in accurately with my experiences under mescal.

I wished to ascertain how the subdued and steady electric light would influence vision and passed into the next room. Here the richly coloured shadows, evidently due to the stimulus of the flickering light, were not obtrusive; but I was able to observe that whatever I gazed at showed a tendency to wave or pulsate. The curtains waved to a marked extent. On close inspection I detected a slight amount of real movement, which doubtless increased the coarser imaginary movement, this latter showed a tendency to spread to the walls. At the same time the matting on the floor showed a very rich texture, thick and felted, and seemed to rise in little waves. These effects were clearly produced by the play of heightened shadows on the outskirts of the visual field. At 3:30 A.M. I found that the phenomena were distinctly decreasing, and soon fell asleep. Sleep was apparently peaceful and dreamless, and I rose at the usual hour without any sense of fatigue, although there was a slight headache. A few of the faint visual phenomena with which the experience had commenced still persisted for a few hours. (Ellis 1897, 1541)

Stages of the Vision Experience

Scientists and laymen alike can recognize that the visual aspects of the peyote experience involve various levels of the nervous system. Although different researchers describe the stages in different ways—and some even comment on the "indescribableness" of the actual visions themselves—there appear to be two basic phases to the vision experience. The first phase is one of form constants in which there are relatively simple images in both pattern and coloring. Most people can experience this early phase only with the eyes closed; its effects may even be enhanced by pressing on the eyeballs with the fingers. It is believed that these form constants are the product of peripheral stimuli within the nervous system, and that the retina within the eye itself is probably the creator of the images (Scheibel and Scheibel 1962, 17). These images are ever-changing in shape or form but may be generally classified in three categories: (1) tapestry, grating, lattice- or fret-work, honeycomb, or filigree; (2) tunnels, funnels, cones, or alleys; and (3) spirals. All may be described as geometrical forms, but often they are incomplete. These images are always brightly colored and often with a "dazzling brightness"—although no one color seems to predominate.

This first phase usually produces amazement, awe, interest, and even delight, but it may also cause profound anxiety because the observer enters a totally different and unknown world. For nearly all, however, there tends to be a removal of themselves from the cares and worries of their everyday existence. Louis Lewin reported the experience of a German physician who took mescaline.

> My ideas of space were very unusual. I could see myself from head to foot as well as the sofa on which I was lying. All else was nothing, absolutely empty space. I was on a solitary island floating in the ether. No part of my body was subject to the laws of gravitation. On the other side of the vacuum—the room seemed to be unlimited in space—extremely fantastic figures appeared before my eyes. I was very excited, perspired and shivered, and was kept in a state of ceaseless wonder. I saw endless

passages with beautiful pointed arches, delightfully coloured arabesques, grotesque decorations, divine, sublime, and enchanting in their fantastic splendor. These visions changed in waves and billows, were built, destroyed, and appeared again in endless variations first on one plane and then in three dimensions, at last disappearing in infinity. The sofa-island disappeared; I did not feel my physical self; an ever-increasing feeling of dissolution set in. I was seized with passionate curiosity, great things were about to be unveiled before me. (Quoted in Lewin 1964, 104)

The second phase of the vision experience is much more complex and may be intermixed with the first phase for a considerable period of time. It is as if slowly the imagination seems to overcome the rational conscious state of reality. First there may be only images associated with conscious thought, but this is soon followed by vivid fantasies that may become indistinguishable from reality. At this point one is having what probably could be described as true hallucinations. The appearance of the outer world—the world of physical reality—is altered; objects appear too big or too small, they move when they should be stationary, or they move in jerky or interrupted ways. The world has changed in the spatial-temporal dimensions and has taken on new but intangible characteristics. It is at this stage that all the senses—but particularly the visual—seem to penetrate the whole personality. During this second phase there are no longer just geometrical forms, but scenes and figures that are brilliantly illuminated and colored; these visions swirl before the eyes in constant movement and deep dimension.

> A particularly picturesque scene: an old-fashioned, single-span bridge across the upper reaches of a river. There are large stones at the sides of the stream and a narrow navigable channel in the middle. A very restful picture . . . now there is a sort of close-up of the bridge which is of the old-fashioned type composed of large blocks of stone . . . while looking at the bridge, a pillar has appeared in the center and transformed it into a double-span bridge. . . . The water is quite still and there are sharply delineated shadows in various tints of blue, green and yellow. . . . The water now starts to ripple; it is now drying up and the rocky bed becomes visible. . . . The bridge is getting smaller and fades away and the rocky bed is gradually becoming transformed into a stone road in a moorland scene. (Critchley 1972, 23)

These visions appear to be external to the observer but are only the products of internal stimuli. They may, in many cases, present previously unconscious material that has been either mentally repressed, condensed, or displaced. Such visual images may flash through the minds of the observers for several hours, although to them time has no meaning. William Braden wrote:

> In the darkness of my mind I saw a Technicolor display of weird-looking growths waving about, like plants in a current at the bottom of the sea. There were stalks and sponges and fan-shaped objects: pink and green and purple. I thought I might wander around down there for several centuries, and I wasn't at all sure that I wanted to. I thought about my family, and it didn't seem fair to go away so long and leave them behind—like Rip Van Winkle. What if they needed me for something, and what if they were all dead when I finally came back? (Braden 1967, 235–36)

General Visual Effects

There are nine general visual effects that will be considered: halos, after-images, changes in brightness and saturation, contrasts, size and symmetry, movement, color discrimination, constant change of vision content, and dimensionality.

Halos. During the early stages of the peyote experience, people frequently see colored halos around objects. Havelock Ellis (1902), in an article on the effects of peyote, referred to "the play of shadowy color, and more especially the violet halos which are seen to play around and over objects and constitute, in my own case at all events, the earliest group of color phenomena seen under mescal [peyote]" (67).

After-Images. Peripheral after-images are usually observed and can be readily distinguished from the actual object or subject being seen. They are like shadows of vivid color which merge with, separate, and then seem to actually coexist with the object. These after-images are often quite intense and may alarm some people because of the confusion they cause. Alwyn Knauer and William J.M.A. Maloney (1913), who made an early study of the effects of mescaline, noted that "these peripheral after-images always had a totally objective character, and they were much more material and real than the most vivid hallucinations produced by the mescaline" (433).

Changes in Brightness and Saturation. "The colors were wonderful. They were the colors of the spectrum intensified as though bathed in the fiercest sunlight. No words can give an idea of their intensity or of their ceaseless, persistent motion" (Prentiss and Morgan 1895, 581). In almost every description of the peyote experience, the observer is nearly overwhelmed by the changes and intensities of colors. This hypersensitivity to color may be one of the most remarkable visual aspects of peyote, and for this reason people often speak of their experiences as fantastic "technicolor hallucinations."

Contrasts. Along with changes in brightness of color, there is a great increase in contrast so that the outlines or details of objects become remarkably sharp and well-defined. A volunteer subject in the experiments of Dr. Erich Guttmann was reported to have said,

> At first they were no more than indefinite white clouds moving on a dull black background. Soon they became more clearly defined, and resolved into an interlacing pattern of white lines. The background became a deeper and a richer black, the lines themselves were exquisitely sharp, fine and delicate. I saw a human eye, some faces traced by the lines in their earlier stages, but for the most part the rich black ground was covered with meaningless arabesques, never still, crossing and interweaving in an endless flow. Every line was duplicated, its outline repeated beside it again and again, growing at each repetition fainter, until infinitely multiplied, it receded into eternity. (Quoted in Guttman 1936, 207)

Size and Symmetry. Heinrich Klüver (1966, 38–40) described the effects of peyote on size perception and felt that some of these impressions resulted from the almost continuous movement of the visions. Moreover, some people are greatly impressed by the symmetry of the figures observed, and if something is not symmetrical, it actually seems out of place.

Movement. Stationary objects seem to move and things in motion to stop. It is as if one were observing a walking person in a series of still photographs. Klüver wrote that one saw the moving tip of a burning cigarette not as

> a continuous line or circle as seen under normal conditions in the dark room, but a number of small glowing balls. At the end of the movement I could still see the entire movement as if it were fixed by a number of

glowing balls standing in the air. Then these balls jumped all of a sudden in a great hurry into the glowing end of the cigarette, but always along the path taken by the cigarette. They did not fade, but all of them went along the curve to the terminal point just as if they were connected by a rubber band. Everything was so distinct that I was able to count the glowing balls; one time I counted 16; there was no luminous line between the glowing balls; it was dark . . . the faster the movement, the more a transition from balls into lines and a decrease in the distance from each other. (Quoted in Klüver 1966, 40)

The apparent movement of fixed objects can have a disconcerting effect on the observer because it tends to create a sense of disorientation and discomfort. While under the influence of mescaline and walking through Canterbury Cathedral, Zaehner noted that

the rose window seemed to expand and contract rhythmically, its pattern continually changing meanwhile. The effect was interesting certainly, but seemed to me less beautiful than its normal state. After a short while I found this growing and shrinking annoying. . . . The patterns made by the actual pieces of glass were meanwhile leading a life independent of the life of the figures depicted on them. . . . They were perpetually on the move, forming and reforming, and not remaining still for a moment. (Zaehner 1961, 214–15)

Color Discrimination. Most people feel they can readily distinguish a wide variety of colors while under the influence of peyote, but psychologists have found that the recognition of colors is somewhat impaired, especially in the red-yellow and green-blue areas; that is, the usual perception of color, as measured by hue discrimination tests, may actually be decreased. Nevertheless, the overwhelming impression of the peyote experience is a grand feast of colors—a "furious succession" from one shade or hue to another. A patient of Knauer and Maloney related that

immediately before my open eyes are a vast number of rings, apparently made of extremely fine steel wire, all constantly rotating in the direction of the hands of a clock. . . . The spaces between the wires seem brighter than the wires themselves. Now the wires shine like dim silver in parts. Now a beautiful light violet tint has developed in them. . . . The wires are moving sinuously in groups from below upwards, passing, crossing and recrossing each other, and curving intricately, so as to produce rarely balanced figures, such as are seen in a kaleidoscope. The background to this gorgeous color panorama was first like faintly illuminated ground glass; it is now a silvery tint, and is deepening into a yellow like pure gold. The colors and their shades vary in different parts of the field. The illumination proceeds from a light, or lights moving suddenly from place to place behind the background, and producing wonderful effects of deep shade and bright light contrast, of sudden bursts of light, and equally sudden extinctions.... The wires are now flattening into bands or ribbons, with a suggestion of transverse striation, and colored a gorgeous ultramarine blue, which passes in places into an intense sea green. These bands move rhythmically, in a wavy upward direction, suggesting a slow endless procession of small mosaics, ascending the wall in single files. The whole picture has suddenly receded, the center much more than the sides, and now in a moment, high above me, is a dome of the most beautiful mosaics, a vision of all that is most gorgeous and harmonious in color. The prevailing tint is blue, but the multitude of shades, each of such wonderful individuality, make me feel that hitherto I have been totally ignorant of what the word color really means. (Quoted in Knauer and Maloney 1913, 429–30)

Constant Change of Vision Content. As noted earlier, there occurs a continual change in brightness and saturation of color, as well as in contrast. There is also a remarkable change in the actual content of the visions, whether they be merely geometric designs or elaborate images of animals or human beings. It is as if one were simply experiencing a film of grand dimensions within a theater from which he cannot depart. The content of the visual experience usually can be influenced to some extent by opening or closing the eyes. With one's eyes open, the usual perception of objects is altered through space or light; halos, after-images, and the vividness of things seem to swirl through the senses of the observer. With the eyes closed, or in a darkened room, the visual experience is one in which figures and images come forth in a constantly changing display of form, color, and motion. Most people feel capable of escaping these visions should they become overpowering because they usually disappear when the eyes are opened. S. Weir Mitchell, a British physician who experimented extensively with peyote at the turn of the century, observed that the tints of intense green and red shifted and altered, and soon were seen no more. Here, again, was the wonderful loveliness of swelling clouds of more vivid colours gone before I could name them, and, sometimes rising from the lower field, and very swiftly altering in colour tones from pale purples and rose to greys, with now and then a bar of level green or orange intense as lightning and as momentary.

When I opened my eyes all was gone at once. Closing them I began after a long interval to see for the first time definite objects associated with colours. The stars sparkled and passed away. A white spear of grey stone grew up to huge height, and became a tall, richly finished Gothic tower of very elaborate and definite design, with many rather worn statues standing in the doorways or on stone brackets. As I gazed every projecting angle, cornice, and even the face of the stones at their joinings were by degrees covered or hung with clusters of what seemed to be huge precious stones, but uncut, some being more like masses of transparent fruit. These were green, purple, red, and orange; never clear yellow and never blue. All seemed to possess an interior light, and to give the faintest idea of the perfectly satisfying intensity and purity of these gorgeous colour-fruits is quite beyond my power. All the colours I have ever beheld are dull as compared to these. (Mitchell 1896, 1626)

Dimensionality. Aldous Huxley (1959) commented upon two common objects that greatly affected him while under the influence of mescaline: curtains and his grey flannel trousers. He was deeply impressed by their texture: "the folds of my grey flannel trousers were charged with 'isness.' To what they owed this privileged status, I cannot say. It is, perhaps, because the forms of folded drapery are so strange and dramatic that they catch the eye" (29).

Heinrich Klüver (1966) remarked that "visionary forms may be of two or three dimensions" (27). He goes on to say that "newspapers, pictures, floors, etc., may assume the appearance of relief maps" and that "human faces seem to undergo certain changes; they become more 'expressive,' the features become more sharply defined" (37–38).

Adverse Reactions

Most of the reported peyote experiences have been described as pleasant, or even ecstatic, but others have been recalled only as terrifying ordeals. Some subjects become morbidly suspicious of everyone; others feel that they are being so persecuted that they may even react violently to those around them. Such reactions, as stated earlier, can be a feature of toxic delirium. Other drugs, the best-known of which is alcohol, may also induce similar frightening experiences. *Delirium tremens*, described as an acute psychosis developing from chronic alcoholism, is an extreme case of a toxic reaction resulting in a mood change usually expressed as fear or terror. However, in contrast to the delirium of the peyote experience, that of the alcoholic has a clouding of the consciousness and certain distinct somatic manifestations (Kolb 1977, 642–43). Other aspects of the two experiences are similar, as, for example, a disorientation in time and space, sensory illusions, and difficulty in communication. The frightening ordeal of a bad peyote experience may also create a feeling of approaching death: "I had a series of attacks or paroxysms, which I can only describe by saying that I felt as though I were dying. It was impossible to move, and it seemed almost impossible to breathe" (Ellis 1898, 135).

Yet even the most terrifying peyote experience in which one seems to feel he or she is facing certain death may lead to what is identified as a profound religious experience. This altered state of consciousness so affects a person emotionally that the experience takes on a mystical, or magical, tone. A beautiful example of such an experience is related by John Rave, the Winnebago Native American whose "autobiography" was so well recorded by Paul Radin. He recalled that after taking a large quantity of peyote,

> I suffered a great deal.... I feared that I might do something foolish to myself (if I remained there alone), and I hoped that someone would come and talk to me. Then someone did come and talk to me, but I did not feel better, so I thought I would go inside where the meeting was going on [a meeting of the peyote church].... I went in and sat down. It was very hot and I felt as though I were going to die. I was very thirsty but I feared to ask for water. I thought that I was certainly going to die. I began to totter over.
>
> I died, and my body was moved by another life. It began to move about; to move about and make signs. It was not I and I could not see it. At last it stood up. The regalia—eagle feathers and gourds—these were holy, they said. They also had a large book there (Bible). These my body took and what is contained in that (book) my body saw. (Quoted in Radin 1963, 64–65)

Not only can the bodily effects be unpleasant, but the visual experience may also contain elements that lead to deep anxiety or fear. The subject may see horrible monsters or may encounter scenes that recall or create terrors or tragedies that cannot be distinguished from the realities of the present. The effect may be so threatening that one finds oneself at the threshold of hell: "Suddenly I found myself at the bottom of a black pit, clawing at the sides, attempting to escape. Leering into the pit was Satan, an evil looking creature with the head of a demon and the body of a spider. He was sneering at me and uttering words, 'Alone! Alone! Alone!' The words echoed throughout the pit like a curse, an affliction from which I could not escape. . . . There was no one with whom to share my plight; I was alone in the pit" (National Clearinghouse for Drug Abuse Information 1973, 8).

John Blofeld (1966), while seeking a mystical experience through the use of mescaline, at first encountered a world of horror: "No words can describe the appalling mental torment that continued for well over an hour. All my organs and sensory experiences seemed to be separate units. There was nothing left of me at all, except a sort of disembodied sufferer, conscious of being mad and racked by unprecedented tension. There seemed no hope of being able to escape this torture—certainly for many hours, perhaps forever. Hell itself could hardly be more terrifying" (28). Fortunately, his experience later became pleasant and greatly rewarding.

Zaehner (1961) also experienced a terrifying picture of the "world beyond": "I seemed to be caught like a wasp in the sordid brown treacle of man's anger. I saw a wild black figure chopping off heads, because it was so funny to see them fall. Worst of all, I came upon the 'lost,' squatting, grey-veiled, among grey rocks, 'at the bottom,' unable to communicate, alone beyond despair" (209). Thus, as Klüver (1966) says, "in some individuals the 'ivress divine' . . . is undoubtedly not very pleasurable; in fact, it is rather an 'ivress diabolique'" (55).

The effects—especially the visual—produced by peyote are unforgettable; they have elements of beauty and of heaven. They also contain the ingredients of a hell, of an experience that may produce fear or terror. Most who experience peyote also say that they now know color as they have never known it before. Some also believe they have seen a little of their inner selves. Yet, despite the wonders and the beauties of such experiences, there is always the inevitable potential of going just a little too far from the level of sanity and slipping over the abyss into insanity. Aldous Huxley (1959), though he believed he had witnessed a truly mystical event while under the influence of mescaline, concluded that his experience was "inexpressibly wonderful, wonderful to the point, almost of being terrifying. And suddenly I had an inkling of what it must feel like to be mad" (45).

CHAPTER FIVE

Medicinal Use of Peyote

The prevention and curing of illnesses have always been a concern of humans; all known cultures have had a collection of medicines to overcome the ravages of both illness and injury—and possibly forestall death. Many of these medicines have been derived or obtained from plants and animals, and the search for new and better ones continues. During their long history in the New World, Native Americans have discovered a wide variety of products to serve as a pharmacopoeia, and one of the most treasured substances in this "medicine cabinet" of herbs and roots is peyote.

USE BY NATIVE AMERICANS

Early Native Americans had an array of plant products that they believed were effective in preventing and combating illness. These indigenous people of the New World, like the early people of Asia, used—and revered—hallucinogenic or other psychotomimetic ("mind-mimicking") substances because of their "therapeutic" values. Respect for these plants persists among many Native American groups in the late twentieth century. Dr. Richard Evans Schultes (1969), the noted ethnobotanist from Harvard University, has commented that "primitive cultures, where sickness and death are usually ascribed to a supernatural cause, have long accorded psychoactive plants a high, even sacred, rank in their magic, medicinal, and religious practices, because their ethnopharmacology often values the psychic effects of 'medicine' more than the physiological. . . . We can no longer afford to ignore reports of aboriginal uses merely because they fall beyond the limits of our credence" (245).

There is no distinction between medicine and religion in many Native American cultures. The Kiowa Apaches, for example, believe that the body and mind are not separate, and the strictly biological or medical approach to the treatment of disease simply is without meaning (Bittle 1960, 142). When a disease strikes, the *person* is affected, not just the body or some part of it. Disease and death are the result of direct action of supernatural agents who often operate through witches. To avoid or cure an illness, Native Americans counteract these supernatural influences by calling into action other supernatural powers by means of prayer, fasting, and "sacred" plants. The primary use of sacred plants by most indigenous people in North America has always been religious and at the same time medicinal because religion and medicine have not been separated. So intimate is this relationship between medicine and religion that often the same term refers "both to natural drugs as well as to any practice or fetish or to supernatural beings that helped to cure the diseases" (de Pasquale 1984, 5).

Many Native Americans of South America do not separate religion and medicine either. Folk healers, or *curanderos*, in Peru use the San Pedro cactus (*Echinopsis pachanoi*), which contains mescaline, the principal mind-affecting alkaloid of peyote (Dobkin de Rios 1968a and 1968b; Joralemon 1984). These healers and their patients believe that the San Pedro cactus, through its profound effects upon the mind, is able to combat the supernatural elements that cause diseases. Not only are the patients fed the plant to induce vomiting (to purge the body of impurities), but curanderos also take the substance to gain insight into the nature of their patients' illnesses.

The North American tribes that use peyote believe it has great curative value. Schultes (1940) remarked that "peyote is, without doubt, the most important medicine used among North American Indians at the present time and seems to be replacing other older, but less spectacular, plant remedies. It is used commonly in daily life as a remedial agent, and the peyote ceremonies of almost all tribes of Mexican and American Indians include a definite curing ritual in which the narcotic [peyote] is administered in large doses to the ill" (178).

One of the main uses of peyote even in religious ceremonies clearly is therapeutic. It is interesting to note that among some Native American tribes the word used to designate peyote is the same as for medicine: "azee" (Navajo), "biisung" (Delaware), "puakit" (Comanche), "makan" (Omaha), "o-jay-bee-kee" (Shawnee), "walena" (Taos), and "naw-tai-no-nee" (Kickapoo). Some Native American accounts of the origin of peyotism refer to the curative nature of the plant. Many peyote leaders, in fact, claim to have been converted to the religion because of being cured by the plant rather than as a result of the search for a vision. Quanah Parker, a leader in the early peyote church (see chapter 2), became a "peyote man" because a Yaqui woman treated him with peyote when he was badly gored by a bull in Mexico. She "doctored him with that medicine that gave him back his life" (Brito 1989, 114–15).

Certainly the therapeutic nature of peyote was one of the main reasons for the spread of peyotism within the United States after its introduction from Mexico, for the hallucinatory or visionary nature of the plant makes peyote one of the most potent medicines for the driving away of evil or supernatural influences (see chapter 2). In some instances Native Americans who have left the peyote religion return when illness strikes them or members of their families, and thereafter proclaim the curative values of peyote. John Rave, the Winnebago whose autobiography was carefully recorded by anthropologist Paul Radin (1926) just after the turn of the century, said, "Do not . . . listen to others talking about this medicine, but try it yourself. That is the only way to find out. No other medicine can accomplish what this has done" (185).

Radin gathered several accounts from the Winnebagos of Wisconsin regarding their strong belief in the power and usefulness of peyote as a medicine. John Rave recounted not only the story of his own cure but also that of others.

> Many years ago I had been sick and it looked as if this sickness were going to kill me. I tried all the Winnebago doctors and then I tried all of the white man's medicines, all were of no avail. I thought to myself, "You are doomed. I wonder whether you will be alive next year?" Such were the thoughts that came to me. As soon as I ate the peyote, however, I got over my sickness. After that I was not sick again. . . .
>
> A man named *Black-waterspirit* was having a hemorrhage at about that time and I wanted him to eat the peyote. "Well," said he, "I am not going to live anyhow." "Eat the peyote nevertheless," I said, "and you will get cured." Before that consumptives never were cured and now for the first time one was cured. *Black-waterspirit* is living today and is very well.
>
> Then there was a man named *Walking-priest*. He was very fond of

whisky, chewed, smoked, and gambled. He was also terribly addicted to women. Indeed he did everything bad. I gave him some of the peyote and he gave up all the bad things he was doing. He had a very dangerous disease. He even had murder in his heart. Today he is living a good life. Such is his desire.

Whoever has any evil thoughts, let him eat the peyote and he will lose all his bad habits. It is a cure for everything bad. . . . Come with your disease for this medicine will cure it. Whatever you have, come and eat this medicine and you will have true knowledge once and for all. Learn of this medicine yourself through actual experience. (Quoted in Radin 1926, 182–84)

Some tribes allow peyote to be used as a medicine only during the night-long religious ceremony and under the direction of the Road Chief, whereas other tribes permit anyone to use it at any time to treat the ill. Many believe that peyote must be used only in certain ways or it may produce bad effects. For example, some Native Americans carefully remove the fuzz or hairs from the top of the "button" because they believe they would become blind if the fuzz were to get in their eyes. On the other hand, some of the unpleasant side-effects have been given therapeutic importance. There is a widespread belief that it is good if a person vomits after taking peyote, because this is simply punishment for one's sins, and vomiting rids the body of all its impurities and evil spirits, some of which may be causing ill health.

Native Americans in Mexico also have long used peyote for therapeutic purposes. Dr. Edward Palmer, who conducted extensive botanical explorations in the late nineteenth century, reported that peyote (which he spelled "biote") was used in Mexico as a fever remedy, to increase lactation in nursing mothers, and "as an external application for back pains" (quoted in Bye 1979b, 143). Palmer also commented that peyote was often used with other plants "for complicated diseases, especially to relieve pain and fever, and like opium, to induce a restful sleep" (143). Bennett and Zingg (1935), in their study of the Tarahumara more than half a century ago, wrote that peyote "is considered a cure for many diseases," including bruises, snake bites, and rheumatism. Peyote "is the ultimate cure when all others have failed" (294). Robert Bye (1979a), in a more recent ethnobotanical study of the Tarahumara, stated that "peyote enables the shaman to see better and to aid in the treatment of his patient," as well as to treat "various bites, wounds, burns, and rheumatic pains" (27). The marketplaces in villages and even major cities still have, along with other native herbs and roots, supplies of peyote (although hidden because of the Mexican government's restrictions on its sale since the 1960s). Schultes (1938) noted that the use of peyote for medicinal purposes by rural Mexicans is so well known that the word "peyote" has been made into the verb "empeyotizarse," which means to self-medicate (705–6). Some Mexicans claim that they have found peyote particularly helpful in alleviating hangovers following alcoholic intoxication.

Native Americans in Mexico primarily use peyote to *protect* from sickness; that is, peyote is taken to create a barrier so that the evil witchcraft of an enemy will have no effect. They believe that there are both natural and supernatural causes of illness. A healthy person has a soul that is content within the body, whereas illness can result from the loss of one's soul by either some supernatural being or by carelessness (Bye 1985, 77–78; 1986, 105). The Tarahumaras consider peyote one of their most powerful plants, one that helps maintain a delicate balance. The cactus is periodically "fed" at fiestas and treated with great respect. The shaman thus works to help maintain the souls in a good condition, both inside and outside the body.

Native Americans in the United States, on the other hand, primarily use peyote to treat individuals *after* they have already become ill, to purge the victim of any evil spirit that might be causing the illness (La Barre 1947, 297). In fact, some of them believe that peyote can cure nearly everything: tuberculosis, pneumonia, rheumatism, scarlet fever, venereal disease, diabetes, influenza, colds, hemorrhages, intestinal ills, cramps, fainting spells, bites, cuts, bruises, wounds, snake bite, cancer, skin diseases, alcoholism, insanity, arthritis, blindness, consumption, spasms, headache, broken bones, breast pains, menstrual disorders, corns, constipation, grippe, hiccoughs, *Datura* poisoning, toothache, and "various additional infectious diseases" (Schultes 1940, 179; McLaughlin 1973, 1–2). A partly chewed peyote button may even be used externally on bruises, wounds, or bites. McLain (1968) also reports that dry tops are powdered and packed "into wounds and around aching teeth," and made into a paste "to rub over burned areas of skin as a pain-relieving ointment" (83). Some Native Americans employ peyote the way other people use aspirin: as a tonic to relieve pain and facilitate healing. To others, however, peyote is far more, for it "protects from almost any conceivable disease and disablement" (La Barre 1947, 297). Mount (1987) claims that peyote also has a "positive reputation as a specific for childbirth among unlicensed, spiritual midwives." He stated that he fed peyote tea to his own wife "with the first contractions," and that the child was born happy and healthy (18–19).

Alcoholism, included in the above list, is one of the most serious health problems among Native Americans. Several studies have been made to determine if peyote and the ceremonies of the Native American Church are effective in treating alcoholism. Pascarosa, Futterman, and Halsweig (1976) wrote that the peyote ritual is successful in the treatment of alcoholics, but "not because of a powerful drug. Tested in a laboratory or hospital setting, the psychedelic drugs alone only rarely bring about significant attitudinal and behavioral change." Yet the "orderly, constructive, and stimulating" mixture of the leader, group, and peyote in the ritual session often leads to positive group interaction and introspection (518). They emphasized that Native Americans only "use the drug-induced altered state of consciousness for insight and exchange during their sessions, and in no way do their daily life patterns require the use of the drug. Their attraction to peyote and the ritual is similar to the patient's reliance on the analyst and does not constitute an addiction in any sense of the word" (523).

Many Western-trained physicians do not believe that peyote is the physical or medicinal panacea that Native Americans believe it to be. Its reputation, in their view, probably arose more from the psychoactive properties of peyote, mainly due to mescaline, and its psychological benefits. On the other hand, one cannot easily dismiss the possibility that one, or several, of the many alkaloids in peyote may have some curative benefits.

EARLY USES BY ANGLO-AMERICANS

At the end of the nineteenth century, peyote aroused considerable interest in medical circles. Some physicians became quite enthusiastic about the possible medical value of *Anhalonium* (an incorrect generic name for peyote; see chapter 8 and appendix A), especially when used in combination with the various narcotics and stimulants widely employed in medicines of that day. One physician, Dr. S. F. Landry (1889), in an article published in the *Therapeutic Gazette*, described its medical benefits and hailed *Anhalonium* as "the best concentrated cardiac tonic we possess" (16). He further suggested it especially for those with respiratory ailments:

> Thinking that it would make a good substitute for nux or ignatia [both containing strychnine and obtained from the genus *Strychnos*] as a respiratory stimulant, I have employed the following mixture of fluid extracts:
> Rx Anhalonium, gtt. i; [one drop]
> Digitalis, gtt. i; [one drop]
> Belladonna, gtt. i; [one drop]
> Cannabis indica [marijuana], gtt. ii; [two drops]
> Water, ss. [one-half ounce]
> This was administered to a phthisical [consumptive] patient at bedtime. After the incipient elevation of temperature the breathing became easy, free, and deep. Next morning the patient was better than for months previous thereto. (Landry 1889, 16)

Realistically, the less than one-fifteenth of one gram of fluid extract of *Anhalonium* specified by Landry probably would have been inadequate to cause any noticeable effect.

A few years later, another doctor enthusiastically reported the results of his use of *Anhalonium* in the *New York Medical Journal*: "In my opinion, anhalonium is a superior cardiac tonic, and, like nitroglycerin, its effects are prolonged after the administration of the drug is withdrawn" (Richardson 1896, 194). He went on to describe in detail how he treated a fifty-year-old patient plagued with cephalalgia (headache) and neuralgia (acute pain of the nerves).

> The patient was given four drops of the tincture every night on retiring, and this dose was not increased. On the evening of the fourth day the neuralgia had left him, being replaced by a peculiar prickling sensation over the surface of the body, but exaggerated in the fingers and toes, and also on the soles of the feet. This sensation lasted but a short time, was unaccompanied by any rise of the temperature of the body or change in the character or rate of the pulse; neither was it especially a source of discomfort to the patient, but quite otherwise. After one week this sensation was experienced for an hour or two after taking the drug, but soon ceased and disappeared altogether in the third week. At this time he suffered a very short but severe attack of frontal cephalalgia, which he endured for about eighteen hours, and for exactly two months since he has been totally free from neuralgia. (194)

At the conclusion of his report, Dr. Richardson remarked that *Anhalonium* might, in the long run, be safer than certain other drugs being tried as substitutes for morphine. "There has for a long time been in use a combination of phenacetine, cannabis indica, caffeine, and camphor monobromide as a succedaneum for morphine. Anhalonium in certain cases will supplant such formulae, and thus avoid the depression of the heart caused by the use of coal-tar derivatives" (195).

Two extensive early medical reports on the possible therapeutic value of peyote were by Drs. D. W. Prentiss and Francis P. Morgan in 1895 and 1896. The latter paper summarized many of their findings and opinions.

> The use of the mescal buttons [peyote] as an antispasmodic in the conditions mentioned—in abdominal pain due to colic or griping of the intestines, irritable cough due to tickling in the throat, nervous headache, and as a substitute for opium and chloral in delirium, great nervous irritability and restlessness, and in insomnia due to pain—is directly in the line of its physiological action. We found in all of our experiments that the first noticeable effect of the drug was a marked sedative and quieting effect upon the muscular system, and we believed this effect to be due to a sedative action of the drug upon the nervous system.
>
> In addition to the uses mentioned, we would suggest the advisability of the use of mescal buttons as an antispasmodic in other conditions in which this class of remedies is indicated, such as hysterical manifestations, spasmodic asthma, convulsions.
>
> The fact that a sense of well-being, mental exhilaration, and happy contentment follows the use of the drug, even in small doses, leads us to believe that it may prove of value as a cerebral stimulant in the treatment of depressed conditions of the mind, such as melancholia, hypochondriasis, and in some cases of neurasthenia. It possesses an advantage over opium in that its use is not followed by the unpleasant effects which often attend the use of that drug. . . .
>
> The conditions, then, in which it seems probable that the use of mescal buttons will produce beneficial results, are the following: In general "nervousness," nervous headache, nervous irritable cough, abdominal pain due to colic or griping of the intestines, hysterical manifestations, and in other similar affections where an antispasmodic is indicated; as a cerebral stimulant in depressed conditions of the mind (hypochondriasis, melancholia, and allied conditions); as a substitute for opium and chloral in conditions of great nervous irritability or restlessness, active delirium

and mania, and in insomnia caused by pain; in color-blindness. (Prentiss and Morgan 1896, 6)

At the end of their paper Prentiss and Morgan discussed the types of preparations that could be made of *Anhalonium*.

> The following preparations may be used:
> *Tinctura Anhalonii* (10 per cent.). Dose, one to two teaspoonfuls (4:00–8.00 Gm.).
> *Extractum Anhalonii Fluidum* (100 per cent.). Dose, ten to fifteen drops (0.50–1.00 Gm.).
> *Pulvis Anhalonii*. Dose, 0.50–1.00 Gm. (seven to fifteen grains). . . .
>
> The taste of these liquid preparations is somewhat disagreeable unless it be disguised by a suitable vehicle, such as a mixture of fluid extract licorice and elixir yerba santa. The powdered drug is best administered in wafer paper, cachets, or capsules. (6–7)

Although these physicians expressed much enthusiasm for peyote, many others in the medical profession refrained from using the plant to any extent, if at all, preferring instead to continue employing the more familiar narcotics and stimulants. For several years, however, editions of the *United States Dispensatory* (a major drug manual) included a brief résumé of peyote and its suggested medical uses under the heading "*Anhalonium lewinii*." About 1920 the entry was deleted and there has been no mention of peyote or its extracts since. The early medical use of peyote by Western physicians was clearly experimental and with inadequate controls. In most cases the dosages were insufficient to produce even mild effects. One must therefore conclude that these early reports were inaccurate, at least with respect to any actual medical changes produced by mescaline.

Possible Antibiotic Actions

Despite the fact that the modern medical profession does not use peyote therapeutically, many Native Americans persist in their belief that it can do much that synthetic, drugstore-type medicines cannot. The widespread claims of therapeutic value by both Native Americans and early physicians encouraged scientists to search for a substance or substances in peyote that might be medically significant. These studies have resulted in the isolation and identification of several naturally occurring alkaloids that have physiological actions similar to other medically used alkaloids, such as strychnine and morphine.

Another series of investigations dealt with possible antibiotic actions of peyote. James McCleary and his colleagues at California State University, Fullerton, studied the effects of peyote extracts on *in vitro* cultures of eighteen penicillin-resistant strains of the bacteria *Staphylococcus aureus* (McCleary, Sypherd, and Walkington 1960, 247–49). All strains tested were inhibited by a water-soluble, crystalline substance that had been extracted from peyote. It was named "peyocactin" by the McCleary group, but its actual identity was not determined at the time. The same research team also tested the *in vivo* effectiveness of "peyocactin" on mice that had been inoculated with toxic strains of *Staphylococcus aureus*; their results showed a definite inhibitory action on the bacteria. They also found other cacti that inhibited at least some of the bacterial strains tested, but none were as broadly effective as peyote (McCleary and Walkington 1964). "Peyocactin" was later identified by G. Subba Rao (1970) of the National Institutes of Health as the alkaloid hordenine (N, N-dimethyl-hydroxyphenylethylamine). This substance is known to have a weak activity similar to epinephrine and is antiseptic as well. Jerry McLaughlin and A. G. Paul (1966) concluded that "peyocactin" (hordenine) has a definite *in vitro antiseptic* action against a wide spectrum of microorganisms (325). On the other hand, their attempt to duplicate the *in vivo* mice experiments of McCleary and his associates did not yield the same results.

Psychiatric Uses

For a number of years scientists in various fields have sought drugs that would effectively treat mental illness, one of the Western world's major medical problems. Despite intensive and well-designed research, treatment of the mentally ill is still frustratingly unsuccessful in a high percentage of cases.

Researchers began to study peyote and its psychotomimetic alkaloid mescaline in the 1920s because of the remarkable psychotic-like state that it produced in normal subjects (Joachimoglu and Keeser 1924, 1107–13; Joel 1929) (see chapter 6 for a discussion of the effects of such drugs as mescaline on the conscious state). Research on the possible therapeutic application of mescaline to psychiatry and psychotherapy continued into the 1970s. Among the numerous, often-contradictory reports that were published on this subject, some held out hope that mescaline might be effective where other drugs or treatments had failed. Unfortunately, most projects led to frustration and disappointment rather than to success. Tranquilizers and antidepressants have become the most widely accepted drugs for psychotherapy, but some researchers believe that the mode of action of mescaline and LSD may still help lead them to answers concerning the causes of mental illness.

The Use of Psychotomimetic Drugs in Psychotherapy

Two of the hallucinogenic or psychotomimetic substances used in psychotherapy include mescaline (the primary psychoactive alkaloid of peyote) and LSD (lysergic acid diethylamide). Some researchers believed these substances were capable of breaking down the defenses of the various "compartments" within the minds of the mentally disturbed, thus permitting both the physician and the patient to discover on a conscious level those behavioral attitudes and responses which were either hidden, reversed, or completely repressed. Through the use of these drugs, "not only is the perception of the outside world affected, but perception of the subject's own personality is also transformed" (Schultes and Hofmann 1979, 176). When a person is under the influence of psychotomimetics, the perception of the outside world and one's ego, normally separated, become blurred. Some people experience psychiatric problems because they are enmeshed in an ego-centered dilemma; they are unable to relate to their surrounding environment because the "I-Thou" barrier is so great. The psychotomimetics were thought to enhance therapy by relaxing this barrier, thus enabling psychiatrists to work more effectively with their patients.

Psychiatrist Walter Frederking (1955) of Hamburg, Germany, stated that mescaline and LSD cause patients to experience "an exhilarating feeling of liberation. Thus, the state of intoxication proves itself to be a phase of the healing process" (263). The same year, psychiatrists Herman C. B. Denber and Sidney Merlis (1955) of Manhattan State Hospital commented that "mescaline releases the repressing forces of the unconscious and permits a free flow of conflictual material hitherto kept out of the awareness. The synthesis of feeling and thoughts results in resolution of the conflict" (468).

Therapists differed in their estimates of mescaline's value, and some expressed alarm that there was too much danger associated with its use. Some felt that the greatest danger was the possibility that a psychotomimetic substance might precipitate a more serious psychotic condition than the one being treated. Others were concerned that the drugs tended "to impair the establishment and maintenance of rapport with the therapist" (Cattell 1954, 242–43), whereas others argued that one of the greatest advantages of these drugs was that *better* patient-therapist relationships were established (Cohen 1967, 581–82).

Mescaline was used extensively in some clinics in the 1960s, but it was largely replaced in the next decade by LSD. Clinicians observed that mescaline and similar drugs generally produced a greater spontaneity of emotional expression, more openness to the experience, less distance in the interpersonal relations between patient and therapist, and a deeper sense of meaning and purpose in life. Mescaline also seemed to heighten patients' emotions, at least temporarily, and in so doing may have lowered many of their defenses; the drug made the "unconscious conscious" and caused patients to relive certain traumas that arose out of past experiences and to which they had not correctly adjusted. For at least some mentally ill persons, these drugs tended to loosen associations and create intense experiences, which, in turn, permitted patients and therapists to deal together with some of the major causative factors of the illness.

Psychotomimetics were used for the treatment of alcoholism, psychosis, suicidal depression, homosexuality, frigidity and other sexual problems, neuroses, phobias and compulsions, psychopathy, and paranoid schizophrenia (Ditman 1968, 46). Therapists claimed at least some improvement of patients having such disorders.

Mescaline was used in two different kinds of therapy. The first is referred to as the small-dose technique, which consisted of weekly or biweekly sessions. Following each treatment the effects of the drug were discussed in group sessions and through visual presentations, such as paintings or drawings. This technique, used mostly in Europe, was also called "psycholysis," the term "lysis" referring to "the dissolving of psychological tensions and conflicts" (Schultes and Hofmann 1979, 178). One report from Germany claimed "a success quotient of about sixty-three per cent.... Nineteen cases out of fifty-four either recovered or were greatly improved; seventeen cases showed good improvement" (Leuner 1963, 67).

The second, or higher-dose, technique was sometimes referred to as "psychedelic therapy." Used mostly in the United States, this technique called upon the use of a single high dose of the drug and fewer sessions, the intent being to produce a powerful state of ecstasy from which the therapist could then begin to restructure the patient's personality (Schultes and Hofmann 1979, 178). According to some therapists, this method actually reduced the total number of hours that they spent with patients. The purpose of the second method was to produce an experience with a tremendous impact that could change patients' outlooks on themselves and life in general. The risk in using this method was that the therapist had to be ready to "rescue" the patient if he or she became too anxious or terror-stricken while under the influence of the drug. Fortunately, an injection of chlorpromazine or another major tranquilizer almost immediately counteracts the effects of mescaline.

The success rate of psychotherapy using mescaline is difficult to evaluate, but most claims of success were with neurotics rather than psychotics. Unfortunately, patients with severe symptoms of schizophrenia did not respond to this group of drugs as well as hoped. Most psychiatrists and psychologists feel that the early claims for success due to mescaline and other hallucinogens actually exceeded their effectiveness. Although admitting that the drugs caused more than simple "placebo responses" in which medication was administered without actual pharmacological action, critics suggested that at least some of the therapeutic effects of the hallucinogens were due to the patients' emotional desires for successful treatment, as well as for closer relationships with their therapists.

Mescaline research continued into the 1960s, but the availability of LSD, and subsequent studies primarily involving this synthetic alkaloid, eliminated the clinical rationale for using peyote or mescaline. Since the prohibition of the use of psychotomimetics by federal law, there have been few reported uses of peyote or mescaline in modern Western medicine.

A Comparison of Schizophrenia and the Effects of Mescaline

Mescaline and the related psychotomimetics, LSD and psilocybin, were believed by some to be "experimental analogs" that produced a clinical picture with major likenesses to the true schizophrenic state. In other words, mescaline and similar drugs could create temporary models of psychopathological syndromes similar to those of true psychoses.

Early in the 1950s, Humphrey Osmond and John Smythies (1952) published an article suggesting a new approach to the study of schizophrenia. They presented the thesis that "mescaline reproduces every single major symptom of acute schizophrenia [including catatonia and thought disorder], although not always to the same degree" (311–12). When comparing the psychological effects of schizophrenia with the mescaline-induced state, they found that both conditions produce sensory disorders (vision, hearing, body image, etc.), motor disorders (catatonia), behavior disorders (negativism, withdrawal, violence), thought disorders (disturbed association, pressure, etc.), disorders of interpretation (paranoid ideas, heightened significance of objects, ideas of influence), delusions, splitting, depersonalization, and mood disorders (fear and terror, depression, euphoria, indifference and apathy, etc.) (312).

Leo E. Hollister (1962) of the Veterans Administration voiced the views of several psychiatrists who felt that some researchers were misunderstanding the concept of "model." He suggested that the term "model psychosis" was misleading if it implied an "actual model of schizophrenic reactions" (87). Evidence has accumulated that mescaline does not create a precise model of schizophrenia and that there are several significant differences between a normal person under the influence of mescaline and one in a schizophrenic state (Hollister 1962, 82–83; Davison 1976, 110–11).

Native Americans of both Mexico and the United States believe that peyote is a potent and important medicine, but the curative nature of the plant probably is due more to their response to its psychoactive properties than to the presence of any true healing chemical substances. Medical opinions concerning the therapeutic value of peyote or mescaline vary widely and have even been contradictory, but it now appears that other drugs, such as tranquilizers and antidepressants, are more effective in reducing the symptoms and abnormal behavior of mental illness, especially as adjuncts to psychotherapy. After failing to produce spectacular successes in either therapy or research, mescaline and related psychotomimetics are no longer valued as medical research tools.

Pharmacology of Peyote

The peyote experience produces profound changes in the sense perceptions that seem to be similar among all peyote users. The human central nervous system is significantly affected by the plant, a matter that has intrigued chemists, biologists, and pharmacologists for more than a century. How does peyote act upon the human body so that some believe it is divine while others feel it is a tool of the devil? How does it alter the body's metabolic processes to produce its remarkable somatic and psychic effects? Although much research has been done, many questions remain unanswered, and there are disputes and controversies over proposed explanations. This chapter deals with some of the physiological aspects of peyote and some of its alkaloids; it also includes a brief discussion of dosages, toxicity, and cytogenetic effects. Mescaline is then compared to some of the other well-known psychotomimetics.

Physiological Effects

Peyote contains more than fifty-five alkaloids and related compounds, many of which probably exist in sufficient quantity to affect human physiology (see chapter 7). However, few have been studied to determine their effects on either humans or other animals. The following alkaloids from peyote have been studied, and some of their physiological actions noted, by Heffter (1894, 78–79); Kloesel (1958, 312), Brossi, Schenker, and Leimgruber (1964); and Kapadia and Fayez (1970, 1717–20).

Lophophorine

This substance is quite toxic, and in doses of about 12 milligrams per kilogram of body weight, it causes violent tetanic convulsions in rabbits. Its action seems to be similar to that of strychnine. In humans, smaller doses cause a strong sickening feeling in the back of the head, a slight decrease in pulse rate, and a hotness and blushing of the face.

Anhalodine

Anhalodine also appears to be a stimulant of the central nervous system but is much less potent than lophophorine and pellotine.

Anhalonidine

This substance seems to produce effects similar to pellotine (see below). In the frog it produces a narcosis followed by a phase of excitability. Large doses produce complete paralysis.

Anhalonine

This drug produces a temporary and incomplete paralysis followed by hyperexcitability in rabbits. The lethal dose in rabbits is 160 to 200 milligrams per kilogram of body weight.

Hordenine

First isolated from barley (*Hordeum vulgare*), this alkaloid has an effect on heart muscle similar to ephedrine. It also causes paralysis of the central nervous system in frogs. Large doses result in hypertension and an accelerated pulse. Very large doses are lethal through the stopping of respiration. Hordenine is also highly antiseptic (see chapter 5).

Pellotine

Eight to 10 milligrams of this alkaloid produce tetanic-like convulsions in frogs. Subcutaneous dosages of about 50 milligrams result in a feeling of drowsiness and a disinclination for all physical or mental effort in humans. It also causes a slowing of the heartbeat and a decrease in blood pressure. This alkaloid can be considered hypnotic in its effects.

Mescaline

One of the major peyote alkaloids, mescaline has been the focus of extensive clinical investigation, and its action on bodily functions has been described by numerous researchers, including Feigen and Alles (1955, 172–75); Deniker (1957, 428–31); Ludwig and Levine (1966, 24); and Kapadia and Fayez (1970, 1718–19). The following is a summary of mescaline's basic somatic effects.

1. Slight increase in blood pressure and pulse rate. Changes in blood pressure are not necessarily correlated with dosage levels.
2. Strong increase in patellar (knee) reflex, especially with doses above 150 milligrams. The peak response occurs in about eighty minutes. High doses produce what can best be described as a "flail-like" reflex.
3. Mydriasis or excessive dilation of the pupils.
4. Evidence of postural instability and some disorders in gait.
5. General increase in motor activity exemplified by fidgeting, etc.
6. Immediate perspiring.
7. Increase in the frequency and amplitude of respiration, often leading to rapid breathing (polypnea).
8. Lowered body temperature (hypothermia) for about the first four hours, followed by a moderately high temperature (hyperthermia).
9. Rapid rise in blood sugar (hyperglycemia), reaching a maximum in about an hour, followed by a return to the initial level in two to four hours.
10. Consistent decrease in blood potassium reaching its lowest point in thirty to sixty minutes.
11. Increased urinary excretion and often a strong desire to defecate.
12. Marked increase in the production of leucocytes (leucocytosis), reaching a peak in two to four hours, followed by a slow decline to the initial level.
13. A flattening of the waves and a general blocking of the alpha rhythm on the electroencephalogram at the time of intense visual perceptions. This probably is due in part to the attention set of the subject.
14. Flushing of the skin, often accompanied by shivering and chills with goose pimples; high doses may cause piloerection (the erection of hairs).

15. Increased salivation.
16. Sensations of hot and cold.

These somatic effects, as well as the accompanying psychic manifestations, may be induced by an oral dose of 5 milligrams of mescaline per kilogram of body weight.

Many of the physiological effects just described can be observed in several other drug-induced states such as those produced by antimalarials, heavy metals, bromides, belladonna alkaloids, cardiac glycosides, and numerous sympathomimetic agents (Jarvik 1970a, 194). They comprise part of the "acute brain syndrome," a drug-induced form of delirium (Kolb 1977, 184).

Physiological Action of Mescaline

Despite the fact that much research has been conducted and many papers published, the mode of action of mescaline and other psychotomimetic substances is only now becoming clear. Giarman and Freedman (1965, 10) suggested that mescaline might exert its effect upon the body by interfering with enzyme systems involved in some of the metabolic processes of the organism, such as those affecting glucose transport into the brain, phosphate metabolism, respiratory reactions, and oxidative phosphorylation by mitochondrial systems within the brain. Supporting data are absent.

Data which best describe the site of action of mescaline involve the biochemical systems that are unique to the brain and nervous system: the chemical reactions that occur in the process of synaptic transmission. One hypothesis has been that mescaline may alter the metabolism of acetylcholine in the brain (Giarman and Freedman 1965, 19). Proper functioning of the nerve endings requires that there be a reciprocal relationship between adrenergic inhibition and cholinergic excitation. Studies have shown that mescaline may alter the storage or receptor sites of acetylcholine; this would result in an increased release of bound acetylcholine and/or a reduced uptake of newly synthesized acetylcholine. The effect of such an imbalance of acetylcholine would be a distortion of synaptic transmissions either in the direction of excessive stimulation or of excessive inhibition.

More-recent research has supported the concept that mescaline may in some way affect the action of serotonin, one of the main neurotransmitters within the central nervous system. Psychotomimetic substances such as mescaline and LSD may either stimulate or inhibit the serotonin (5-HT) receptors in the brain (Browne and Ho 1975, 429–30; Hollister 1982, 34). Rech and Commissaris (1982), McCall (1986), and Jacobs (1987) believe that mescaline and the other indole and phenylethylamine psychotomimetics act as agonists, either directly or indirectly, on central nervous system serotonin receptors by occupying their binding sites. The primary effect probably is upon postsynaptic receptors. McCall (1986) noted that even low doses of mescaline potentiate for several hours the effect of serotonin and norepinephrine. Thus, if the effect of the psychotomimetics "occurs in sensory pathways in the central nervous system, then a mechanism of receptor sensitization, in distinction to disinhibition, might account for the altered perceptual reactivity produced by these drugs" (362).

Jacobs (1987) has concluded that these psychotomimetic substances act at a postsynaptic serotonin-receptor site, specifically of the 5-HT_2 subtype. He further commented that "under normal conditions serotonin exerts an equal action at all of its receptor sites. However, when one site [such as at one of the 5-HT_2 subtype] is selectively activated or blocked, whether by drugs, endogenous biological factors, or environmental toxins, it leads to a perturbation which may be manifested as anxiety, migraines, or hallucinations" (390–91).

Some investigators question whether mescaline is directly active as the hallucinogenic agent; rather, they suggest that a psychoactive substance is synthesized from mescaline. Significantly, 60 to 90 percent of the original dose of mescaline is excreted in the subject's urine in its original form, most of the remainder being eliminated by oxidative deamination as 3,4,5-trimethoxyphenylacetic acid (TMPA) (Ott, 1976, 114; Lemberger and Rubin 1976, 67). Small amounts of 3,4,5-trimethoxybenzoicacid, N-acetylmescaline, and N-acetyl-3,4-dimethoxy-5-hydroxyphenylethylamine have also been identified in small amounts in urine. None are psychoactive (Baselt and Cravey 1989, 506). Because LSD and mescaline have similar effects on the central nervous system, it has been suggested that mescaline may become modified so that its psychoactive conformation resembles a part of the LSD molecule, thus enabling both substances to act on the same receptor site in the central nervous system (Nichols et al. 1977; Aghajanian 1982, 95). This could be accomplished if the amide side chain of mescaline were "to fold back to either position 2 or 6 with hydrogen bonding, thus forming approximately the indole ring of LSD" (Fujimori and Alpers 1971, 399). Although it is possible that the psychoactive nature of mescaline may be due to a metabolite, further data are needed to settle the matter.

Dosages

Native Americans who participate in the all-night peyote ceremony may ingest between three and more than a dozen plant tops or "peyote buttons." Arthur Heffter (1898, 404), a nineteenth-century German pharmacologist, estimated that there were 4.6 to 6.8 grams of mescaline in every kilogram of dried peyote. However, Fischer (1958) calculated that only about 3 percent of the dried plant may be mescaline, and this quantity can be obtained only if the material is finely ground prior to extraction. He estimated that less than 1 percent of the mescaline in a peyote button could be obtained by chewing and swallowing, meaning the average participant in a peyote ritual who eats five to ten peyote buttons might get up to 0.2 gram (200 milligrams) of mescaline. Thirty buttons, which would be considered a very large dose, would have little more than half a gram (500 milligrams) of mescaline.

The dosages of pure mescaline sulfate or mescaline hydrochloride, the usual forms into which mescaline is extracted or synthesized, vary with the method of application. Oral doses are usually about 5 milligrams per kilogram of body weight, or a total of 300 to 500 milligrams (Jarvik 1970a, 195; Shulgin 1979b). Intravenous administration of mescaline is 2 to 4 milligrams per kilogram of body weight (Aghajanian 1982, 95).

Ruth Underhill (1952, 148) reported that humans have taken 8 grams of mescaline powder—more than ten times the usual dose—without apparent toxic reactions.

Toxicity of Mescaline

A major concern of researchers, physicians, and therapists is the toxic level of mescaline. Some of the early anti-peyote reports by representatives of the Bureau of Indian Affairs claimed that the deaths of several Native Americans were at least hastened by use of the plant during religious ceremonies (Newberne 1925, 15). There are no recorded cases of deaths due directly to the eating of peyote; however, Dr. Chris Lawrence, a physician in New Mexico, reported the death of a young Native American who drank a cup of peyote "tea" and within three hours suffered cardiopulmonary arrest and died. The subsequent autopsy showed severe effects of alcoholism, but very low levels of mescaline in his blood and urine. Dr. Lawrence also said the pathologist concluded that "death is probably best explained by chronic alcoholism with alcoholic hepatitis with a contributing factor of blood loss from a Mallory Weiss tear, resulting from violent vomiting induced by ingestion of peyote" (pers. comm.).

There is one reported case of a fatality due to the ingestion of an overdose of mescaline, in which an individual in California climbed a steep hill near the ocean and leaped off the edge as if attempting "a swan dive," falling 200 meters (600 feet) to his death. An autopsy showed high levels of mescaline in his blood, liver, and urine (Reynolds and Jindrich 1985).

Medical practitioners sometimes have difficulty identifying the symptoms of overdoses of peyote or mescaline. Because a diagnosis will often reveal symptoms of "acute-gastroenteritis, acute abdomen, acute pancreatitis, or other serious illness, none of which are really present," Teitelbaum and Wingeleth recommend that if the patient has "nausea, vomiting, sweating, cramps, abdominal pain and hallucinations," an immediate urine analysis should be carried out before commencing "aggressive medical or surgical therapy" (37).

The lethal dose of mescaline in human beings must be extrapolated from data obtained on laboratory animals. Louise B. Speck (1957, 79) of the University of Colorado determined that the lethal dose (LD50) in rats is 370 milligrams per kilogram of body weight when administered intraperitoneally. She observed that flexor convulsions were followed by cardiac arrest that resulted in death. A dose of 411 milligrams per kilogram of body weight (equivalent to 1.32 mM of mescaline base per kilogram) caused death in less than thirty minutes. Several drugs, such as insulin, the barbiturates, and physostigmine, when given in high doses along with mescaline, seemed to increase the toxicity of mescaline.

Subsequent studies by Hardman, Haavik, and Seevers (1973, 301–2) disagreed with those of Dr. Speck. The Hardman team found the LD50 value in the rat to be only 132 milligrams per kilogram of body weight, or 0.53 mM of mescaline base per kilogram when administered intraperitoneally. They also determined the LD50 values for four other laboratory animals: the mouse (212 mg/kg or 0.86 mM/kg of base), the guinea pig (328 mg/kg or 1.33 mM/kg of base), the dog (54 mg/kg or 0.22 mM/kg of base administered intravenously), and the monkey (130 mg/kg or 0.53 mM/kg of base administered intravenously). The rat is more than twice as sensitive to mescaline as either the mouse or guinea pig. The dog, on the other hand, is by far the most sensitive species studied; moreover, it exhibited the characteristic psychic and somatic effects of mescaline more consistently than did the other animals.

It should be noted that the LD50 of the dog is about ten times the usual dose taken by humans. However, it should be emphasized that all organisms clearly do not have the same level of resistance to mescaline.

TOLERANCE

Tolerance is the reduction in an organism's intensity of reaction to a continued certain dose of a drug, thus requiring ever-increasing amounts to cause the desired reaction (Fujimori and Alpers 1971, 364; Wyatt et al. 1976, 466–76). The true narcotics, such as opium and cocaine, produce a high degree of tolerance, and often the concept is included in the definition of narcotic (see chapter 9). Experiments with laboratory animals have produced conflicting data with regard to mescaline tolerance. Appel and Freedman (1968, 273) found that in tests conducted with the rat, tolerance developed in two to three days using daily doses of 10 milligrams of mescaline per kilogram of body weight. The behavioral action measured was the pressing of a bar for a food reward. These researchers found that the rats developed tolerance to several of the psychotomimetic drugs and that the rate at which tolerance developed was inversely related to the relative potencies of the drug. Therefore, mescaline, the least-potent psychotomimetic of those tested, caused the most rapid development of tolerance. Other researchers found that mescaline tolerance rapidly developed in tests involving the conditioned avoidance response (Smythies, Sykes, and Lord, 1966, 436). Contrary results have been obtained by some other investigators, however; Bridger, Mandel, and Stoff (1973) found that mescaline "proved to be excitatory even when a high dosage was chronically administered" (132).

Perhaps some of the disagreement among investigators has arisen because tolerance seems to develop in research animals for certain effects of mescaline, but not for others. Yet evidence is contradictory here as well. Fischer (1958, 391), for example, reported that in dogs, a complete tolerance developed to some of the somatic effects, such as vomiting, whereas no tolerance arose for the psychic aspects. Jacobsen (1963, 488), on the other hand, claimed that tolerance for the vegetative (somatic) phase did *not* develop to the same degree as for the psychic phase. He also observed that although tolerance developed rapidly, it regressed with equal speed following the withdrawal of mescaline doses, and the initial sensitivity of the organism was regained within just three days.

Humans rapidly develop a tolerance to both the psychic and somatic effects of mescaline (Wyatt et al. 1976, 467), but this tolerance is rapidly lost when the person is withdrawn from the drug (Jacobsen 1963, 488).

Cross-tolerance, the condition in which an individual becomes tolerant to the usual effects of a given dose of one drug and shows the same tolerance to a substituted drug, has been reported between mescaline and LSD, psilocybin, DOM (2,5-dimethoxy-4-methylamphetamine) (Wyatt et al. 1976, 467), DMM (N,N-dimethylmescaline), and DMPE (3,4-dimethoxyphenylethylamine) (Smythies, Sykes, and Lord 1966, 438–43). Cross-tolerance strongly suggests that all of these substances may have the same, or similar, final pathway or mechanism of action (Fujimori and Alpers 1971, 364). Wyatt and colleagues (1976, 467) reported that the most complete cross-tolerance is to mescaline in humans who are LSD-tolerant.

CYTOGENETIC EFFECTS

A team of researchers first reported possible human genetic damage by LSD in 1971 (Dishotsky et al. 1971). Four years later a group of California physicians reported the results of the first study of the effects of peyote on human cells (Dorrance, Janiger, and Teplitz 1975). Their cytogenetic studies involved the Huichol people of northern Mexico, a group that has long used peyote ceremonially and medicinally (see chapters 1 and 5). Some Huichols eat peyote ceremonially as many as thirty-five times a year, with consumption beginning as early as age six and continuing throughout their lives. Some Huichol groups do not use peyote and thus served as the experimental controls. The data clearly showed that "no serious chromosome damage had occurred" in the peyote-using Huichols (Dorrance, Janiger, and Teplitz 1975, 301). Future research should involve other groups of peyote-using people, but the Dorrance team's findings indicate that peyote does not adversely affect human hereditary materials.

British scientists at the University of Aberdeen (Harrisson, Page, and Keir 1976) reported that mescaline is an extremely potent inhibitor of the formation of the spindle apparatus in cell division, or mitosis. The intriguing aspects of these studies are that they were performed on *in vitro* human cells, and the effects were more pronounced and less toxic than with the better-known spindle inhibitors colchicine and its methyl derivative colcemid.

Mescaline, LSD, Psilocybin, and Marijuana

Mescaline (derived from peyote), LSD (a synthetic ergot alkaloid derivative), and psilocybin (obtained from certain fungi of the family Agaricaceae) are remarkably similar in the effects they produce, but they do differ chemically (fig. 6.1). Although the experiences one has with each substance have only a few qualitative differences, the quantities needed to produce the same psychic and somatic levels of action differ greatly. An effective oral dose of LSD is 50 to 200 micrograms (µg), whereas 300 to 500 milligrams (mg) of mescaline are needed to produce a comparable reaction (Fujimori and Alpers 1971, 363). The average dose of psilocybin is 20 to 40 mg (G. Fisher 1965, 154–56). Ludwig and Levine (1966, 21) found the effective clinical dose of mescaline to be 5,000 to 10,000 µg per kilogram of body weight, compared to 1 to 2 µg of LSD per kilogram of body weight, or 100 to 200 µg of psilocybin per kilogram. Thus, the potency of LSD is several thousand times that of mescaline.

There are two problems, however, in comparing these three substances. First, the subjective effects of mescaline, LSD, and psilocybin are so similar that researchers have been unable to find a reaction that is distinctly associated with one drug and not another (Jarvik 1970b, 288; Ott 1976, 111). In other words, there is no definitive tag by which the observer of the subject can conclusively identify which of these three substances has been taken (Schwartz 1988, 123). Second, comparative studies of the effects of the three substances on humans indicate that the environmental setting, dose, and other variables may cause even a single subject to have distinctly different reactions to the same drug from one time to another (see chapter 4). The variation of experience may therefore be most easily and logically accounted for by one of these variables *rather than* actual different effects of the three substances.

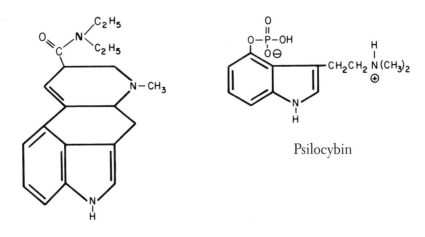

FIGURE 6.1. Chemical structures of LSD and psilocybin. They produce remarkably similar effects to those of peyote, but differ chemically. Compare their chemical structures to that of mescaline (fig. 7.2).

Psychiatrist Walter Frederking (1955, 264) reported in one of the early studies comparing LSD and mescaline that mescaline tended to release stronger emotions and to create a greater sense of anxiety; LSD, on the other hand, seemed to produce more intense physical effects. Wikler (1957, 84) reported that mescaline produced images and experiences that tended to be more expansive and wish-fulfilling than LSD. Hollister and Hartman (1962, 237) studied the effects of these substances on color discrimination and reported that although mescaline and psilocybin increased the number of errors, they did not occur in any specific area of the color range. LSD, on the other hand, significantly increased the number of errors in the red-yellow and yellow-orange areas.

Hoffer and Osmond, in their book titled *The Hallucinogens* (1967), reported a study by Rinkel and his associates, who determined that people under the influence of mescaline who had more-athletic physiques "developed marked physiological disturbances, euphoria, and very few somatic changes. The intellectual (aesthetic) types suffered mental confusion, disruption of mental function, and few physiological changes but many somatic complaints" (41). Both LSD and psilocybin produced similar effects, but they were less pronounced than those induced by mescaline.

Of the reported variations in the effects of the three substances, two factors are consistent enough to be significant. First, instances have been reported in which the effects of LSD recurred as much as eighteen months later without use of the drug again; however, this phenomenon is rare in those who have used mescaline, and has not been reported by those who have taken or studied psilocybin (D. Fisher 1968, 69). Second, the three psychotomimetics vary in the length of the experience they produce. If the drugs are given orally, the effects of mescaline usually appear in two to three hours and may last for twelve or more hours. LSD, on the other hand, tends to be faster-acting, with effects felt in about thirty to forty minutes, and the experience tending to last only eight to twelve hours. Psilocybin acts more rapidly than either of the other two, producing effects in twenty to thirty minutes, and lasting only about four hours (Ludwig and Levine 1966, 21). Although dosage, recurrence of effects, and length of experience are notable distinguishing factors, the differences in effects between mescaline, LSD, and psilocybin appear relatively minor, with the similarities strongly outweighing the differences.

Marijuana

Marijuana, obtained from the genus *Cannabis* of the family Cannabaceae, is another common plant product frequently referred to as a hallucinogen (Ott 1976, 49; Biel, Bopp, and Mitchell 1978, 166). Smoking marijuana produces short-term effects of only a few hours, including euphoria, relaxation, a lessening of inhibitions, an enhancement of sensory perception, and a limited impairment of the more-complex psychomotor reactions (Wells 1973, 29–30). There is no indication that marijuana is addicting, and no evidence that tolerance develops (Grollman 1965, 251). No cross-tolerance is exhibited between marijuana and any of the other psychotomimetics. Marijuana users sometimes experience a "reverse tolerance" (or increased sensitivity) in which the person who has smoked "pot" many times actually requires less than the novice to get an equal reaction. Probably this is due in large part to one's learning how to smoke a marijuana cigarette in order to obtain the greatest amount of Δ^9-THC, the active compound (see fig. 7.3) (Wells 1973, 32).

Most marijuana smokers perceive a distortion of space and time, and some encounter a rapid flight of ideas, but few experience hallucinations. Certainly marijuana is a mind-altering drug, but, in fact, it seems to have little in common with mescaline, LSD, and psilocybin (Grinspoon 1969, 19).

Peyote, or one or more of its alkaloids, has been the subject of numerous investigations by pharmacologists and others interested in its mode of action. The publications that have resulted from this research, however, indicate that much is still not known, and that some of the data available in the 1990s can be interpreted in many ways.

Chemistry of Peyote

Earlier chapters have dealt with how peyote affects the senses and thereby has caused the plant to be admired and revered—but also condemned—by various people. It is now appropriate to consider the chemistry of the alkaloids in peyote, some of which alter sensory functioning.

Alkaloids

Alkaloids were traditionally defined as containing nitrogen, of complex molecular structure, as coming from plants, and as having a marked physiological action on animals (Pelletier 1983, 2). Recent research, however, has shown that the meaning of the term has changed considerably and that older definitions are no longer satisfactory. Some researchers have suggested calling the basic N-heterocyclic compounds such as anhalinine and mescaline (figs. 7.1 and 7.2) "true alkaloids," meaning that a portion of their structure consists of a benzene ring containing at least one element other than carbon, namely nitrogen. Other biogenetically related substances containing nitrogen, including amines, are called "protoalkaloids" (Mothes and Luckner 1985, 15). This term still has problems, especially considering the numerous exceptions to the traditional definition. Various plant bases, such as the betaines and muscarine, would be classified as "protoalkaloids" because they are not N-heterocyclic compounds. Some chemists exclude aliphatic diamino, triamino, and tetra-amino compounds, although their macrocyclic derivatives are included. Some even exclude the phenylalkylamines, including β-phenylethylamine, hordenine, and mescaline (Pelletier 1983, 4).

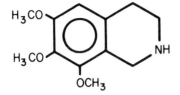

FIGURE 7.1. Anhalinine, one of the alkaloids found in peyote

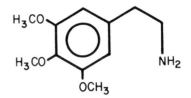

FIGURE 7.2. Mescaline (3,4,5-trimethoxy-β-phenylethylamine) is primarily responsible for the psychoactive properties of peyote. It is not a true indole alkaloid, but rather a phenylethylamine.

A large majority of the 7,000 known alkaloids are basic or alkaline in nature because they contain nitrogen in the form of amine groups (-NH) (the nitrogen atom is directly attached to a carbon), which behave as bases. Some compounds, however, such as colchicine and piperine, are classified as alkaloids but are not basic in nature (Pelletier 1983, 25). About 40 percent of the plant families have at least one alkaloid-bearing species, yet less than 10 percent of some 10,000 plant genera are known to have them (Pelletier 1983, 3). Many alkaloids are also found in animals, bacteria, fungi, and protists. Pelletier (1983) recommends "that the term *alkaloid* be defined on the basis of structure and not on the basis of source or biogenesis" (22–23). He also proposes a simple definition: "An alkaloid is a cyclic organic compound containing nitrogen in a negative oxidation state which is of limited distribution among living organisms" (26).

Apparently, amino acids are the natural precursors of most alkaloids (Mothes and Luckner 1985, 17). Alkaloids normally exist in plants as salts of various acids, and they can be extracted by dilute acids, water, or alcohol (Swan 1967, 2). Alkaloids may be synthesized in different regions of the plant than where they are concentrated; such regions of accumulation include actively growing tissue, epidermal or hypodermal cells, vascular sheaths, and laticifers. They are usually present in vacuoles rather than the cytoplasm of plant cells, but they may also be attached to cell walls. Sometimes they are restricted to special cell types (Robinson 1981, 5).

Alkaloids are familiar to and widely used by humans. In 1805, morphine was the first alkaloid to be isolated by scientists, from extracts of the opium poppy (*Papaver somniferum*). Two other derivatives have also come from the opium poppy: codeine (the methyl ether of morphine) and heroin (the diacetyl derivative of morphine) (Swan 1967, 127–29). Although heroin has been greatly misused, morphine is of considerable medical value. Other well-known alkaloids are nicotine, caffeine, cocaine, quinine, atropine, and strychnine.

It is unclear what functions alkaloids perform within the plant. Traditionally physiologists thought that they were simply reservoirs for nitrogen, or waste products of the plant's metabolic reactions similar to those in animals, such as urea and uric acid. Their presence in vacuoles, which are the sites of waste deposits, supported this idea. However, Trevor Robinson believes that alkaloids are not simply inert compounds but appear to be dynamic, both in total concentration and in rate of turnover (1974, 430–35; 1981, 8). Some alkaloids, such as morphine, show diurnal fluctuations in concentration within the plant, which strongly suggest that they are an active part of the plant's metabolism.

Considerable research has been done on the function of alkaloids in plants (Waller and Nowacki 1978, 143–181; Schlee 1985, 56–64). Some evidence suggests that alkaloids may protect the plant from predators (herbivores) and parasites; however, only a small percentage of alkaloid-bearing plants seem to derive such a benefit. Waller and Nowacki (1978, 164–80) investigated the role of alkaloids in the coevolution of plants and insects, as well as the eventual effects upon herbivorous vertebrates. They concluded that "the insects probably have had a great influence on the establishment of the alkaloid character in some plant families, and the herbivorous mammals are responsible for preserving this character; furthermore, even if the alkaloid, due to simultaneous evolution of insects, can no longer defend the plant from *specialized* vermin, it is still preventing predation by unadapted herbivores" (180). Alkaloids may also serve as plant growth regulators or as substances helping to maintain the correct ionic balance. At present, scientists simply are unsure of the function or functions that alkaloids perform. However, it seems clear that the energy expended by the plant in forming alkaloids would indicate a past—or present—function of some importance. Moreover, if they were simply waste products, they would be accumulated in dead leaves, to be shed, or converted into some other form.

During the second half of the nineteenth century, chemists actively studied alkaloids from numerous plants, but not until chromatographic and spectrophotometric methods became available did the number of known alkaloids rapidly increase. Of the approximately 7,000 alkaloids that have been identified, nearly half were discovered since 1970 (Mothes and Luckner 1985, 16).

Traditionally alkaloids have been extracted with an acidic aqueous solvent that dissolves them in the form of salts. Another method is to make the substance strongly alkaline with sodium bicarbonate; the free bases are then extracted into various organic solvents such as ether or chloroform. Once an acidic aqueous solution is obtained, it is then made basic and treated with an organic solvent so that neutral and acidic water-soluble compounds are left behind. Chromatography can then separate the extract into its pure components.

Alkaloids apparently are produced in several major biosynthetic pathways in plants, being derived from common amino acids and other small molecules through simple reactions such as dehydration, oxidation, decarboxylation, and condensation (Robinson 1967, 259–61). Alkaloids are widely distributed within the plant kingdom, particularly the flowering plants. H. L. Li and J. J. Willaman (1968), in a paper dealing with the distribution of alkaloids in the angiosperms and its phylogenetic implications, came to the interesting conclusion that there is a definite correlation of alkaloid distribution with presently accepted ideas about the phylogeny of the angiosperms based on morphological evidence. There seems to be "a high incidence of alkaloid occurrence in the morphologically primitive Magnoliales-Ranales complex and related groups, a lower incidence in various phylogenetically intermediate groups and a progression of high incidence in all of the phylogenetically advanced but unrelated groups" (251). More recently, Waller and Nowacki (1978) studied the role of alkaloids in chemotaxonomic studies. They concluded that alkaloids "are of little value in the larger taxonomic units, but they can prove to be of great value in smaller ones" (47).

Li and Willaman (1968, 241, 251) surveyed the cactus family and found that 90 percent of those tested contained alkaloids, a remarkably high percentage. However, Jan G. Bruhn (1971, 54) of the University of Uppsala, Sweden, using field tests and laboratory checks, determined that only about 40 percent of the cacti that he tested contained alkaloids. This wide discrepancy is difficult to understand, and further tests are necessary, but it is nevertheless evident that a large number of cacti do contain alkaloids.

Before dealing with the alkaloids of peyote, it is important to consider briefly the relationship of alkaloids to psychotomimetics. Many alkaloids are not psychoactive, and some psychotomimetics are non-nitrogenous compounds. Two types of psychotomimetics are found in the angiosperms and

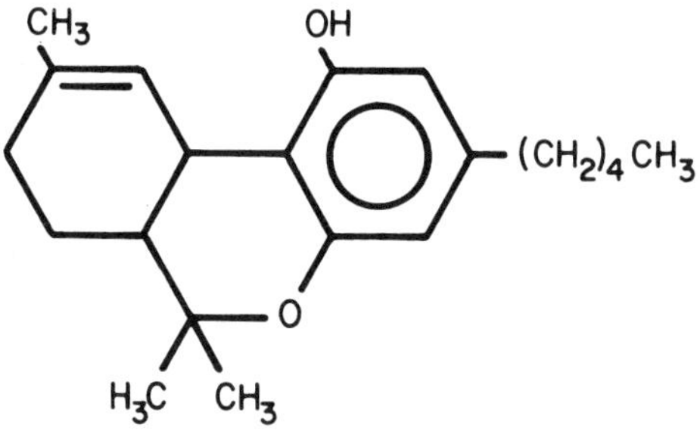

FIGURE 7.3. Δ⁹-Tetrahydrocannabinol (THC), an active non-nitrogenous compound found in marijuana

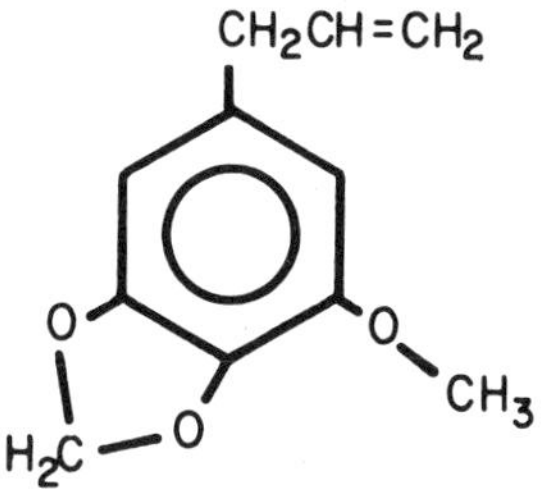

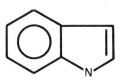

FIGURE 7.5. The basic indole ring, common to most hallucinogens

FIGURE 7.4. Myristicin (3-methoxy-4,5-methylenedioxy), an essential oil found in nutmeg

fungi, the two groups of organisms containing the vast majority of mind-altering chemicals. The first type is composed of nitrogenous compounds that are alkaloids or alkaloid derivatives; most of these probably are derived from the amino acid tryptophan and include the β-carbolines, ergolines, indoles, isoquinolines, isoxazoles, β-phenylethylamines, quinolizidines, tropanes, and tryptamines (Schultes 1970, 571). An example is mescaline from peyote (see fig. 7.2).

The second type of psychotomimetic contains non-nitrogenous compounds, which chemically are dibenzopyrans and phenylpropenes. Within this non-nitrogenous and non-alkaloid group are the active ingredients of marijuana and nutmeg, for example. One of the active non-nitrogenous compounds of marijuana is Δ⁹-tetrahydrocannabinol (fig. 7.3), while one of the active substances in nutmeg is the essential oil myristicin (3-methoxy-4,5-methylenedioxy) (fig. 7.4).

From the standpoint of chemical structure, almost all psychotomimetics

Alkaloids : 137

contain an indole nucleus or ring (fig. 7.5). An important exception, however, is mescaline, the primary psychoactive alkaloid of peyote (see fig. 7.2).

Peyote Alkaloids

In 1888 the German pharmacologist Louis Lewin published the first paper dealing with the chemistry of peyote. The previous year Lewin had obtained dried material of peyote from Parke, Davis, and Company in Detroit, first by mail and then a few months later while visiting the United States (Bruhn and Holmstedt 1974, 358–60). (This latter material was identified by botanist Paul E. Hennings as a new species of peyote [see chapter 8].) Lewin (1888a, 235) was able to extract and isolate a substance which he named anhalonine (based on the generic name *Anhalonium*, widely but incorrectly used for peyote at the time). This material produced no hallucinatory effects and probably was a mixture of several alkaloids.

Lewin's work on peyote stimulated other German pharmacologists to study the plant, and Arthur Heffter (1894, 77–78; 1896), who was able to obtain fresh plant material from European cactus dealers, isolated a second peyote alkaloid which he named "pellotin" (pellotine). Later he received plants from Parke, Davis, and Company, and within a short period of time was able to isolate three additional peyote alkaloids; by self-experimentation he determined that one of these was the main psychoactive substance of peyote, which he named mescaline (Heffter 1898). This substance was the first hallucinogenic compound to be chemically identified. Heffter named the other two alkaloids anhalonidine and lophophorine. E. Kauder (1899, 194) isolated and named an additional peyote alkaloid, anhalamine. Thus, by the end of the nineteenth century, six alkaloids had been isolated from peyote, and researchers were finding alkaloids in other species of cacti as well.

During the first third of the twentieth century relatively little progress was made in understanding the chemistry of peyote, with the exception of the important work by Ernst Späth and his colleagues, beginning in 1918 and continuing until 1939. During this time Späth and his associates published nineteen papers on the chemistry of peyote and other cacti, which included the descriptions of five additional peyote alkaloids: anhalinine, anhalidine, N-methylmescaline, N-acetylmescaline, and O-methylanhalonidine (Kapadia and Fayez 1973, 9). Späth (1919, 129) was also successful in synthesizing mescaline for the first time. Moreover, Späth showed that the peyote alka-

FIGURE 7.6. The chemical structure of a phenol (a phenolic compound), in which the hydrogen atom is replaced by a hydroxy group. Peyote alkaloids such as dopamine are phenolic compounds.

FIGURE 7.7. Toluene is an example of a non-phenolic compound.

loids were primarily of two types based upon their functional groups: the phenolic and non-phenolic alkaloids. A phenolic compound is one in which a hydrogen atom attached to an aromatic ring is replaced by a hydroxyl group (fig. 7.6). A non-phenolic substance, on the other hand, has some other group attached to the carbon, such as methoxy or methyl side chains (fig. 7.7). Within the peyote cactus, dopamine is a phenolic alkaloid (see appendix B, no. 5), whereas mescaline is a non-phenolic alkaloid (see fig. 7.2 and appendix B, no. 17).

Virtually nothing more was done on the chemistry of peyote until about 1965. Since that time several groups of researchers, including Jerry L. McLaughlin and colleagues (1965, 1966, 1967) and Govind J. Kapadia and his coworkers (Kapadia and Fales 1968a, 1968b; Kapadia and Fayez 1970, 1973), have discovered, isolated, and described about forty-five additional chemical substances from peyote. This research activity has been due to the widespread interest in psychotomimetic substances, as well as to the new methods of isolation and identification mentioned earlier.

More than fifty-five peyote alkaloids have been isolated and characterized; some are unique to peyote, whereas others are known to occur in different cacti or even other plant families. Appendix B contains a listing of these peyote substances, their structures, molecular formulas, and other technical data. Only some of the more important and abundant alkaloids are described or discussed in this chapter.

The peyote alkaloids currently are classified on the basis of their ring structure as either phenylethylamines or isoquinolines. The basic structure of the former is that of a benzene ring with an ethylamine side chain, as, for example, β-phenylethylamine (fig. 7.8), which is not only a primary plant con-

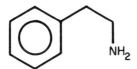

FIGURE 7.8. β-Phenylethylamine (mescaline) has a benzene ring with an ethylamine side chain. It is a primary constituent of peyote and also the product of protein decomposition.

Mescaline + HCHO ⟶ Anhalinine

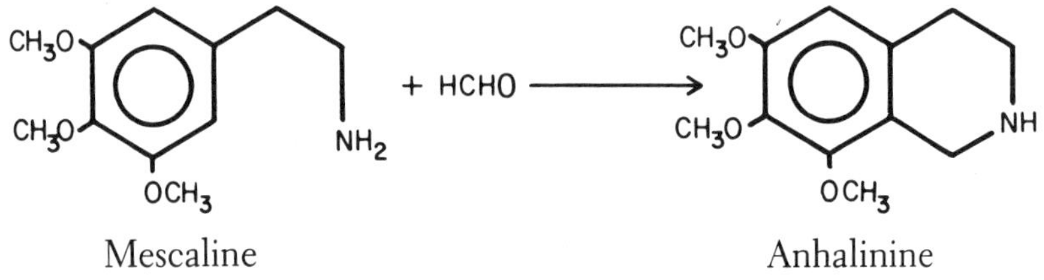

FIGURE 7.9. Mescaline and formaldehyde can react to produce the alkaloid anhalinine.

stituent but also the product of protein decomposition. The isoquinolines can be either simple or complex. Probably the substituted β-phenylethylamines can be considered as precursors of the more simple isoquinolines. An example of a simple isoquinoline alkaloid in peyote is anhalinine (see fig. 7.1).

Phenylethylamines

This group of amine-containing compounds includes epinephrine, tyramine, and hordenine. β-phenylethylamine has been isolated from *Viscum album*, a mistletoe of the family Loranthaceae, and many members of the family Fabaceae (Leguminosae). In this group of compounds, unlike most alkaloids, nitrogen is not part of the heterocyclic system. Biogenetically the phenylethylamines are believed to be related to the naturally occurring aromatic amino acids such as tyrosine and phenylalanine (Schütte and Liebisch 1985, 188).

The most important phenylethylamine alkaloid in peyote is mescaline, or 3,4,5-trimethoxy-β-phenylethylamine (see fig. 7.2). Other phenylethylamines in peyote include various compounds closely related to mescaline, such as tyramine, hordenine, and candicine (Robinson 1981, 25). (See appendix B for a complete listing of the phenylethylamine compounds.)

Isoquinolines

The isoquinoline alkaloids are highly variable: the simple ones have only a single aromatic nucleus, whereas the more complex isoquinolines may have several such nuclei. Apparently peyote and other cacti have only simple isoquinolines. Some of these alkaloids are structurally so similar to the β-phenylethylamines that there must be a biogenetic pathway between the two groups. For example, mescaline and formaldehyde could easily yield anhalinine, a common isoquinoline alkaloid of peyote, through the closing of the side chain and the loss of water (fig. 7.9) (Reti 1954, 7).

There are twenty-three types of isoquinoline alkaloids known to occur naturally in peyote (see appendix B). Some of the more important ones include anhalamine, anhalinine, anhalidine, anhalonidine, pellotine, anhalonine, lophophorine, and peyophorine.

Alkaloidal Amines and Amino Acids

Govind J. Kapadia and M.B.E. Fayez (1970, 1710) brought attention to the fact that little research has been done on the non-basic compounds within peyote and other alkaloid-bearing plants. They characterized succinimide, malimide, maleimide, citrimide, and isocitrimide lactone. Kapadia and colleagues described two rare pyrrole derivatives from peyote named peyoglunal and peyonine (Kapadia and Highet 1968, 191; Kapadia et al. 1970), and they also noted the presence of two mescaline- and dimethylmescaline-related lactams, which they named mescalotam and peyoglutam (Kapadia and Fales 1968a, 1688). Other conjugates with Krebs acids are present in peyote and are characterized briefly in appendix B.

Little has been done to isolate and identify amino acids in peyote, especially the non-proteinic ones. To date four non-proteinic amino acids have been studied: mescaloxylic acid, mescaloruvic acid, peyoruvic acid, and peyoxylic acid (Kapadia and Fayez 1973, 14).

PEYOTE ALKALOIDS IN OTHER CACTI

Many of the alkaloids first identified from peyote have subsequently been found in other members of the cactus family. For example, mescaline occurs in the following genera (Willaman and Schubert 1961, 57–59; Kapadia and Fayez 1970, 1702–4; Agurell et al. 1971, 184–85; Neal et al. 1972; Ma et al. 1986, 735–36):

From South America:
Echinopsis (Trichocereus)
Gymnocalycium
Stetsonia
From North America:
Opuntia
Pachycereus
Pelecyphora
Polaskia
Stenocereus

Peyote alkaloids are found in the following plants:

Tyramine: *Echinopsis (Trichocereus)* spp. (Agurell 1969); *Echinopsis rhodotricha, Stetsonia coryne* (Agurell et al. 1971, 184–85); *Obregonia* (Neal, Sato, and McLaughlin 1971); *Cereus* spp. (Bruhn and Lindgren 1976); and *Espostoa huanucoensis* (Mata, McLaughlin, and Earle 1976)

N-Methylmescaline: *Pelecyphora aselliformis* (Neal et al. 1972)

N-Methyltyramine: *Ariocarpus agavoides* (Bruhn 1975, 16); *Coryphantha* spp. (Bruhn et al. 1975, 202); *Echinopsis (Trichocereus)* spp. (Kapadia and Fayez 1970, 1702–4); *Espostoa huanucoensis* (Mata, McLaughlin, and Earle 1976); *Mammillaria pectinifera* (Bruhn and Bruhn 1973, 248); *Obregonia* (Neal, Sato, and McLaughlin 1971); *Stetsonia coryne* (Agurell et al. 1971, 184–85); and *Turbinicarpus* spp. (West, Vanderveen, and McLaughlin 1974)

Hordenine: Numerous cacti including *Ariocarpus* (Bruhn and Bruhn 1973, 244); *Cereus* (Agurell 1969, 213–14); *Coryphantha* (Hornemann, Neal, and McLaughlin 1972); *Echinopsis (Trichocereus)* (Willaman and Schubert 1961, 59; Agurell 1969, 213–14); *Espostoa* (Mata, McLaughlin, and Earle 1976, 461–63); *Mammillaria* (Bruhn and Bruhn 1973, 244–48); *Obregonia* (Neal, Sato, and McLaughlin 1971); *Pelecyphora* (Neal et al. 1972); and *Turbinicarpus* (Agurell et al. 1971, 184–85; Bruhn and Bruhn 1973, 245)

Anhalidine: *Pelecyphora aselliformis* (Neal et al. 1972) and *Stetsonia coryne* (Agurell et al. 1971, 184–85)

N,N-Dimethyl-3-hydroxy-4,5-dimethoxyphenylethylamine: *Pelecyphora aselliformis* (Neal et al. 1972) and *Ariocarpus agavoides* (Bruhn and Bruhn 1973, 244)

Anhalonidine: *Trichocereus pachanoi* and *Stetsonia coryne* (Agurell et al. 1971, 184–85)

Pellotine: *Pelecyphora aselliformis* (Neal et al. 1972)

Anhalonine: *Echinopsis (Trichocereus) terscheckii* and *Gymnocalycium gibbosum* (Willaman and Schubert 1961, 59)

4-Hydroxy-3-methoxyphenylethylamine: *Carnegeia gigantea* (Bruhn and Lundstrom 1976)

3,4-Dimethoxyphenylethylamine: *Pelecyphora aselliformis* (Neal et al. 1972); *Pachycereus pecten-aboriginum* (Bruhn and Lindgren 1976); and *Carnegeia gigantea* (Bruhn and Lundstrom 1976)

Chemically Different Species of Peyote

Much of the early work on the chemistry of peyote involved plant material that was poorly documented. Frequently the site and/or time of collection of the plants were unknown, and in some cases improperly identified plant material was used. It is well known that in many alkaloid-producing plants, considerable chemical variation can occur from one population or geographic locality to another. Therefore, James S. Todd of Whitman College and I collected and carefully documented peyote material during a period of two weeks from throughout its geographical range; he then undertook a systematic analysis of the presence and relative amounts of some of the major peyote alkaloids in the plants collected. He also compared the relative concentrations of these alkaloids in the tops and in the roots. Several interesting results were obtained.

1. Hordenine, though present in *Lophophora williamsii* roots, was not found in *L. diffusa* (see chapter 8 for the botanical characteristics of these two species).
2. A "slight but definite inverse relationship exists between mescaline and pellotine with respect to the amount present in the various populations" (Todd 1969, 397). Todd suggested that these two

alkaloids might be formed from a common precursor via competitive pathways, which apparently is true for the first few steps (see figure 7.15). He also noted that both mescaline and anhalinine are "essentially absent" from *L. diffusa*.

3. There are chemical variations from one population of peyote to another in three Mexican populations. Todd's population A represented *L. williamsii* from near Monclova, Coahuila; population B was *L. williamsii* from El Huizache junction in the state of San Luis Potosí; and population C was *L. diffusa* from north of Vizarrón in Querétaro. The two populations of *L. williamsii* are quite similar chemically, whereas *L. diffusa* totally lacks anhalinine, anhalonine, and hordenine. It has lesser concentrations of all the other alkaloids studied (lophophorine, mescaline, anhalamine, and anhalonidine), with the notable exception of pellotine, which is in a much greater concentration. Todd also found that with few exceptions (such as hordenine), there was relatively little variation in alkaloid concentration between roots and tops. His research demonstrates the importance of having correctly identified and fully documented plant material for chemical studies.

Jan G. Bruhn and Bo Holmstedt (1974, 371–85) verified much of Todd's research and added additional information to support the conclusion that the two species of peyote differ significantly in chemical content. Their studies also clarify to a considerable extent the confusion that has long existed with regard to the identity of plant materials studied in the last century by Lewin, Heffter, and others. They believe that specimens from the southern population (*L. diffusa*) were misidentified as those from the north (*L. williamsii*) because of insufficient field data and an absence of botanical knowledge about the two groups. This aspect of peyote history is discussed further in chapter 8 and in appendix A.

Mescaline

Mescaline, or 3,4,5-trimethoxy-β-phenylethylamine (see fig. 7.2), is the alkaloid primarily responsible for the psychotomimetic action of peyote. Appendix B lists several analogs of this alkaloid. As stated earlier, mescaline is not a true indole alkaloid but rather a phenylethylamine. It is a colorless, strongly alkaline oil or an oily crystalline material having a melting point of 30°C to 32°C. It is soluble in water, alcohol, and chloroform, but much less soluble in ether. One of the most common salts, mescaline hydrogen sulfate, is often the form in which mescaline is isolated from plant materials because it is insoluble in alcohol and almost insoluble in cold water (but highly soluble in hot water). It forms very distinct and brilliant prisms, and has a melting point of 183°C to 186°C. Most medical experimenters have used either mescaline hydrogen sulfate or mescaline hydrochloride because of their solubility and relative ease of handling.

There is some disagreement regarding the amount of mescaline in peyote. Kapadia and Fayez (1973, 15) and Reti (1953, 324) claimed that it is about 6 percent. Kelsey (1959, 232), on the other hand, contended that the concentration of mescaline in "the dried cactus" is only 0.9 percent, or 360 milligrams of the alkaloid for 40 grams of dry plant material. Bergman (1971, 696) reported unpublished data by M. H. Seevers of the University of Michigan indicating that 3 grams of dried peyote contain 45 milligrams of mescaline (1.5 percent). Probably the correct concentration of mescaline for most peyote plants would be about 1 percent of the whole dried plant (Crosby and McLaughlin 1973, 416).

Echinopsis pachanoi, formerly in the genus *Trichocereus* and known as the San Pedro cactus, is used extensively by Native Americans in Peru for "medicinal purposes" (see chapter 5). It contains about the same quantity of mescaline as does peyote (Crosby and McLaughlin 1973, 416; Joralemon 1984, 402). *Echinopsis* mescaline is usually taken orally as a brew made by slicing young stems as one would cut a loaf of bread, placing the pieces in a kettle with about five gallons of water, and boiling the contents for seven hours (Sharon 1972, 119–20). Undoubtedly this would greatly concentrate the amount of mescaline. Other species of *Echinopsis* have about one-tenth as much mescaline as the San Pedro cactus (Shulgin 1979a, 42).

Pelecyphora aselliformis, one of the other North American cacti containing mescaline, has only a very minute amount, approximately 10^{-5} percent of the dry weight of the plant (Shulgin 1979a, 42). This cactus is called *peyotillo*, or "little peyote," by Native Americans and is reputedly used medicinally (Neal et al. 1972, 1133); however, it is difficult to see how enough *Pelecyphora* could be ingested to produce psychotomimetic effects. An interesting question has been posed by David Sands, a research biologist at the University of Montana, who wonders why this plant would retain the necessary enzymes to produce such a small quantity of mescaline. He suggests that perhaps some sort of metabolic "emergency" might induce the plant to produce much larger quantities (pers. comm.). It is also possible that in *Pelecyphora*, mescaline might be an intermediate product in a process that normally converts nearly all the intermediate substances.

Several methods are available to isolate and identify mescaline within plant or animal tissue. Extraction is accomplished by methanol; this initial stage is then completed by filtration of the extract and its evaporation to dryness. The extract is next treated with chloroform and 0.05 N hydrochloric acid in a separatory funnel; the aqueous portion is retained after several washings with chloroform. Ammonia or sodium carbonate is added to the aqueous solution in sufficient quantities to produce a slightly basic solution with a pH of about 8. This is followed by further extraction with chloroform and chloroform-ethanol (3:1). After adjusting the pH to 10, a final chloroform-ethanol extraction is made. The chloroform extract that contains the alkaloids is then dried. The alkaloids can be separated into the phenolic and non-phenolic groups by passing the extract (redissolved in chloroform) through Amberlite IRA-400 (OH^-) ion-exchange resin (Agurell 1969, 208–10). If thin-layer chromatography is used for alkaloid separation and identification, several spray reagents are particularly useful. For example, Fluorescamine (4-phenylspiro [furan-2 (3H), 1′-phthalan]-3,3-dione) readily distinguishes the phenylethylamines from the tetrahydroisoquinolines (Ranieri and McLaughlin 1975). Mescaline may then be identified by comparison with known samples using spectrophotometry.

"Applied chemists" within the drug cult have devised ingenious methods of extracting pure mescaline from dried or fresh plant material (Superweed 1968, 7–12). The basic process varies somewhat, but a typical one is as follows: the plant material is first boiled to extract the alkaloids; this extract is then made basic by the addition of sodium hydroxide (lye). Next benzene is added to further separate the alkaloids. The aqueous and benzene portions are allowed to separate following a gentle shaking. Dilute sulfuric acid is next added in small quantities to the benzene portion and the solution again shaken. The mixture is allowed to stand, and the process is repeated several more times with the addition of more dilute sulfuric acid each time. A white precipitate will soon settle and can easily be dried. This is mescaline sulfate, and further steps can make it quite pure.

As mentioned earlier, the first laboratory synthesis of mescaline was accomplished by Ernst Späth (1919), who took 3,4,5-trimethoxybenzoic acid, transformed it into its corresponding aldehyde, produced nitrostyrene by reaction with nitromethane in an ethanol solution containing alkali, then reduced the nitrostyrene with zinc dust and acetic acid to the corresponding oxime, which was finally reduced by sodium amalgam to mescaline. His process yielded slightly more than 20 percent mescaline. Numerous modifications of the Späth technique have been published, and many other approaches have likewise been described. Perhaps the most fruitful variations of the synthesis of mescaline originated with Ramirez and Burger (1950), who found that reduction of the nitrostyrene side chain could be accomplished in one step having a high yield by using lithium aluminum hydride. Other workers soon found that the same reduction could be accomplished by electrolysis.

Whereas the method described by Späth involved a two-step reduction of the nitrostyrene having a yield of only 24.9 percent, the same reduction utilizing lithium aluminum hydride results in a yield of 89 percent. A completely novel and relatively simple approach to mescaline synthesis was presented by Tsao (1951), starting with gallic acid and again utilizing lithium aluminum hydride to reduce the methyl ester and nitrile. His synthesis is gallic acid → 3,4,5-trimethoxybenzoic acid → methyl ester of 3,4,5-trimethoxybenzoic acid → 3,4,5-trimethoxybenzyl alcohol → 3,4,5-trimethoxybenzyl chloride → 3,4,5-trimethoxyphenylacetonitrile → mescaline.

Structure-Activity Relationship of Mescaline

Mescaline (3,4,5-trimethoxy-β-phenylethylamine) is structurally similar to the neurotransmitter epinephrine and to amphetamine (figs. 7.10, 7.11, and 7.12). Biochemists and pharmacologists have determined that increasing the number of oxygen-linked substituents on the benzene ring increases the psychoactive properties of compounds structurally similar to mescaline; however, both the position of the substitution on the ring and the nature of the substituent seem to be important. For example, a 3,4 compound is inactive; this is also the case for a mono-substituted compound. Oxygen-linked substituents must be present in at least three positions (2,4,5 or 3,4,5, for example) for psychoactivity.

The addition of more methoxy groups produces an even more active substance (Smythies et al. 1967, 383). Five methoxy groups in the compound 2,3,4,5,6-pentamethoxyphenylethylamine is eight times more potent than

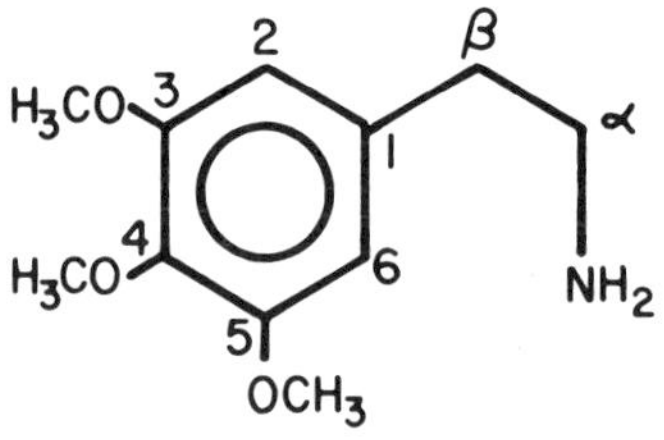

FIGURE 7.10. The structure of mescaline, showing the numbering of the carbon atoms of the benzene ring and the Greek lettering of the carbon atoms of the side chain.

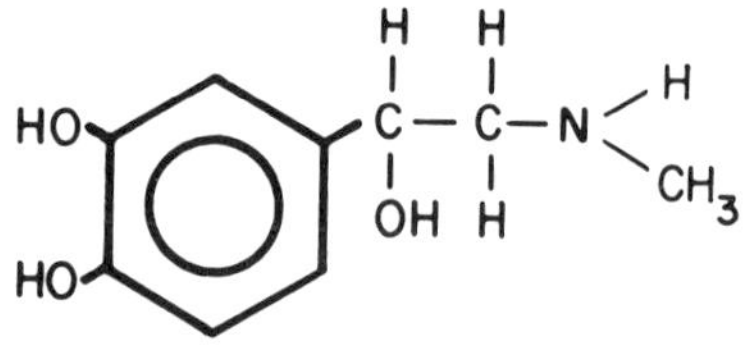

FIGURE 7.11. The neurohormone epinephrine is also chemically similar to mescaline.

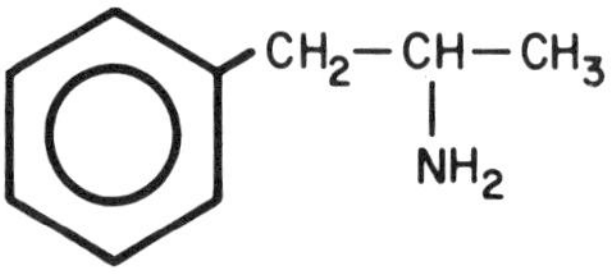

FIGURE 7.12. Amphetamine is structurally similar to mescaline.

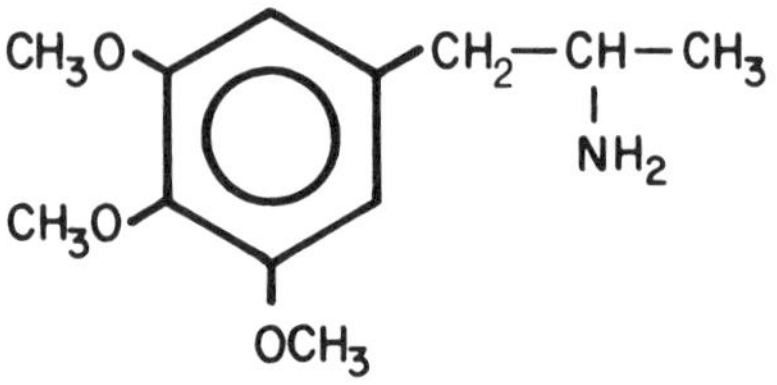

FIGURE 7.13. 3,4,5-Trimethoxy-amphetamine has more than twice the potency of mescaline.

mescaline. A methoxy group (H_3CO-) can even be replaced by a hydroxy group (-OH) and some activity will be present (Hollister 1968, 23–24). If the methoxy group on the number 5 carbon atom is removed, the potency of mescaline is reduced by half (Glennon, Kier, and Shulgin 1979). If the methoxy group of the number 4 carbon atom is replaced with a hydroxy group, activity is eliminated (Smythies 1963, 18). However, the substitution of a methyl, halo, or alkoxy group for the same methoxy group greatly increases activity (Shulgin 1978, 272), and the substitution of iodine (I) for a methyl group at the number 4 position increased the activity by more than 40 times (Glennon, Kier, and Shulgin 1979, 907).

Experiments involving the length of the side chain demonstrate that three carbons are optimum (Shulgin 1970, 36). A homolog of mescaline, 3,4,5-trimethoxyamphetamine (3,4,5-trimethoxyphenylisopropylamine or TMA) (fig. 7.13), differs only in the nature of the side chain but has more than twice the potency of mescaline (Shulgin 1973, 52). Kang and Green (1970, 64) showed that psychotomimetic activity is not only produced by the presence of the aromatic nature of the benzene ring, but also by the presence of an amino group in the side chain. If an ethyl group (CH_3CH_2--) is added to the α carbon of the side chain, the molecule loses its capability of causing psychic disturbances or any activity in the central nervous system. The addition of a methyl group to the side chain at the α position (see fig. 7.10) greatly increases its psychoactivity, but if it is added to the β position it is inactive (Nichols 1981, 842).

Biosynthesis of Peyote Alkaloids

Chemists have long attempted to unravel the mysteries of the biogenesis of natural plant products, especially some of the mind-altering alkaloids. Workers at the turn of the century hypothesized several such biogenetic routes based solely on the chemical structures of some possible precursors and the structure of the final compound itself. It was not until the 1960s, however, that an appropriate tool for confirming these hypotheses was developed: the radioactive tracer. The administration of labeled precursors, often carbon-14, is now a routine technique in the biosynthetic studies of alkaloids. A possible precursor is labeled and introduced into the normal metabolism of the plant; after the regular biosynthetic processes have proceeded for a period of time, the alkaloids are extracted and degraded by carefully controlled processes in order to determine the positions of the labeled atoms within the structure of the alkaloid, and any intermediate compounds that may still exist. Interestingly, many early hypotheses concerning possible biogenetic pathways of peyote alkaloids have been confirmed or modified only slightly by the use of labeled substrates, but not all alkaloid origins have yet been determined.

Reti (1950, 268) proposed a scheme in which the well-known aromatic amino acids phenylalanine, tyrosine, and 3,4-dihydroxyphenylalanine served as the precursors and were converted step-by-step into the more complex phenylethylamines and tetrahydroisoquinolines by decarboxylation, oxidation, and methylation—reactions that are well known in all living organisms. Reti's proposal has been largely supported by more-recent work; researchers have found that phenylalanine produces dopamine via tyrosine and tyramine, which, in turn, is O-methylated to give 3-methoxy-4-hydroxyphenylethylamine. This is then hydroxylated to 3-methoxy-4,5-dihydroxyphenylethylamine, which is O-methylated to give 3,5-dimethoxy-4-hydroxyphenylamine. O-methylation of this compound produces mescaline (see fig. 7.15) (Schütte and Liebisch 1985, 193). This summary, however, does not tell the whole story, because the biosynthesis of other important alkaloids in peyote has interesting relationships to the biosynthesis of mescaline.

Biosynthesis of Mescaline and Hordenine

Reti (1950, 268) proposed that tyrosine was the precursor of mescaline; others suggested that phenylalanine formed hordenine via tyrosine by simple stepwise methylation (Schütte and Liebisch 1985, 191–92). Primarily as a result of the experiments of Edward Leete at the University of Minnesota, Stig Agurell and his colleagues at the Royal Pharmaceutical Institute in Stockholm, and Jerry L. McLaughlin and his colleagues at Purdue University, the biosynthetic pathways of these two important peyote alkaloids have been determined with a relatively high degree of assurance. There are at least two routes by which mescaline may be synthesized, but apparently only one pathway for the manufacture of hordenine. Tyrosine, which is formed from phenylalanine, can be regarded as the main precursor for these two alkaloids (figs. 7.14 and 7.15).

The biosynthesis of mescaline may follow two possible biogenetic routes from tyrosine (Kapadia and Fayez 1970, 1712–14) as shown in figure 7.15:

1. Tyramine may be created from tyrosine by demethylation, then converted to dopamine which, by one of two different routes, is changed into 3-methoxy-4,5-dihydroxyphenylethylamine. The more common route is via nor-mescaline (3,4,5-trihydroxy-β-phenylethylamine), whereas the alternate pathway is through the formation of 3-methoxy-4-hydroxyphenylethylamine. Mescaline (3,4,5-trimethoxy-β-phenylethylamine) is the final product and is formed from 3-methoxy-4,5-dihydroxyphenylethylamine with a single intermediate compound, which is 3,5-dimethoxy-4-hydroxyphenylethylamine.
2. Tyrosine is converted to dopa which is then changed to dopamine. Again, 3-methoxy-4,5-dihydroxyphenylethylamine can then be formed in either of the two ways described in step one.

Hordenine is also synthesized from tyrosine by the tyramine route. However, the tyramine in this route is converted to N-methyltyramine. A relatively simple additional methylation would then produce hordenine.

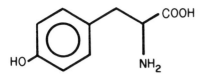

FIGURE 7.14. Tyrosine can be regarded as the main precursor of mescaline.

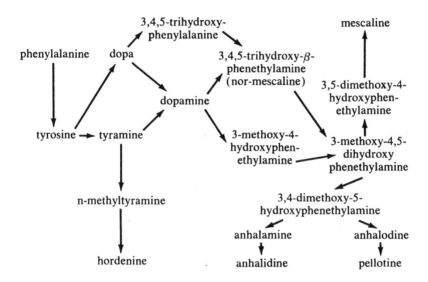

FIGURE 7.15. Probable biosynthetic pathways of some of the phenylethylamine alkaloids of peyote.

Biosynthesis of the Tetrahydroisoquinolines

Reti and other biochemists contended that, like the phenylethylamine alkaloids, those of the tetrahydroisoquinoline group also come from some of the naturally occurring aromatic amino acids, such as phenylalanine and tyrosine, through such basic reactions as decarboxylation, oxidation, and ring closure (see fig. 7.15). Because these biosynthetic pathways are numerous and highly complex, several problems still remain in our understanding of the natural formation of the many tetrahydroisoquinoline alkaloids (Kapadia and Fayez 1970, 1714–17). Current research indicates that, indeed, tyrosine and dopamine are the probable precursors. The pathway from tyrosine involves the production of dopa by hydroxylation, followed by decarboxylation to dopamine. This compound is converted to 3-methoxy-4-hydroxyphenylethylamine and then to 3-methoxy-4,5-dihydroxyphenylethylamine. A change in methyl group orientation results in 3,4-dimethoxy-5-hydroxyphenylethylamine. This compound, in turn, can be converted into such tetrahydroisoquinolines as anhalamine, anhalidine, anhalodine, anhalonidine, and pellotine (Schütte and Liebisch 1985, 192, 195–96). Anhalonidine produces pellotine by simple N-methylation and vice versa. There is a distinct possibility that a form of mescaline (3-dimethylmescaline) may also serve as a precursor for some of the tetrahydroisoquinolines, such as anhalamine and anhalonidine, by passing through such intermediate forms as peyoruvic acid and peyoxilic acid.

The peyote cactus contains the greatest variety of alkaloids known to occur in the cactus family. Some of them are found only in this genus, while others, such as hordenine and mescaline, are in several groups. These compounds present a fascinating challenge to biochemists who are attempting to determine their biogenesis within the plant. Pharmacologists are likewise interested in these alkaloids because of their profound physiological effects on humans and other animals (see chapter 6). Botanists also find the natural alkaloids of interest, for they apparently indicate certain relationships and evolutionary trends within plants (Waller and Nowacki 1978, 1–47). As the next chapter makes clear, alkaloids have certainly been important in the present botanical classification of peyote.

Chapter Eight

Botany of Peyote

Until the 1950s there had been no careful and extensive botanical studies of peyote. Plants which were brought into the chemist's laboratory—or the horticulturist's or cactus collector's greenhouse—had little or no documentation concerning the site or date of collection. Even the name of the collector often was unknown, and there was little knowledge about the ecology of the plants. As a result, plants from the same population were sometimes given different scientific names, and those of separate regions often were given the same name.

This absence of botanical understanding, primarily due to insufficient field and laboratory studies, consequently resulted in mistakes and confusion for historians, anthropologists, chemists, pharmacologists, and others. For example, numerous references have been made to peyote as belonging to the genus *Anhalonium*. This name is botanically invalid because it was applied to the group of cacti that had earlier been named *Ariocarpus*; hence, the later name *Anhalonium* cannot be used for that group or any other group of plants such as peyote. However, the name *Anhalonium* had been employed so widely for about a century that few people other than botanists specializing in taxonomy and nomenclature were aware that the name should not have been used. The resulting confusion and difficulties can probably never be completely straightened out.

Virtually all of the early chemical and pharmacological studies of peyote were based on poorly documented and incorrectly identified specimens. The German toxicologist Louis Lewin obtained material from Parke, Davis and Company of Detroit in 1887, but the exact natural source of the plants is unknown. Lewin extracted a highly toxic substance from the plants, which he named "anhalonine" (Lewin 1888a and 1888b), but later study showed it to be a mixture of alkaloids (Bruhn and Holdmstedt 1974, 360).

In 1891 Arthur Heffter, a German chemist, received 42 kilograms of peyote from Mexico (Bruhn 1977, 29). These plants were to be the basis of some of the most important—and confusing—pioneer chemical studies of peyote. Heffter discovered that the plants he had received belonged to two distinct groups based on the alkaloids present, but he was unable to distinguish the groups on structural or morphological grounds. Since he had no collection and field data, he decided that peyote simply consisted of two chemical forms. Bruhn and Holmstedt (1974, 384–85) researched the literature dealing with this period of peyote history, and they concluded that Heffter's batch of plants actually consisted of the two distinct species of peyote, which do have definite alkaloid differences (see chapter 7). A better botanical understanding of the group, as well as proper scientific data, would have prevented the introduction of much confusing information into the literature that has persisted for more than a century.

This frustrating botanical chaos concerning peyote existed for so long that in the 1950s Gordon A. Alles, a research biochemist interested in peyote's psychotomimetic properties, personally financed my graduate studies so that I could determine the botanical relationships of the group and unscramble the nomenclature. The botanical aspects are now much clearer.

The peyote cactus is a flowering plant of the family Cactaceae, a group of about 100 genera and perhaps 1,500 species; most are fleshy, spiny plants found primarily in the dry regions of the New World. Some of the common characteristics of cacti are not readily evident in peyote, except for the obvious one of succulence. Spines, for example, are present only in very young seedlings. However, the cactus areole, a highly specialized axillary bud that produces flowers and spines (which are modified leaves), is well pronounced in peyote and is identified by a tuft of hairs, or trichomes. Flowers arise from young areoles within the center of the plant, and like other cacti, the perianth segments of peyote flowers are not sharply divided into sepals and petals; instead there is a transition from small, scale-like outer perianth parts to large, colored, petal-like inner ones.

Another characteristic that shows peyote belongs in the cactus family is the absence of visible leaves in both juvenile and mature plants. Leaves are greatly reduced and only microscopic in size; even the seed leaves, or cotyledons, are almost invisible in young seedlings because they are rounded, united, and quite small. Also, the vascular system of peyote is like that of other succulent cacti in which the secondary xylem is very simple and has only helical wall thickenings.

Botanical History

Peyote was first described by Europeans in 1560 (de Sahagún 1938, 3:118), but it was not until the nineteenth century that any plants reached the Old World for scientific study. Apparently the French botanist Charles Lemaire was the first to publish a botanical name for peyote, but unfortunately the name that Lemaire used for the plant, *Echinocactus williamsii*, appeared in the year 1845 without description and only in a horticultural catalog. Prince Salm-Dyck (1845), another European botanist, provided the necessary description to botanically validate Lemaire's binomial, but no illustration accompanied either the Lemaire name or the description by Salm-Dyck. It was not until 1847 that the first picture of peyote appeared in *Curtis's Botanical Magazine* (Hooker 1847, plate 4296) (fig. 8.1).

In the second half of the nineteenth century the characteristics and scope of the large genus *Echinocactus* were disputed by several European and American botanists. Gradually its limits were narrowed and new genera were proposed to contain species that had once been included in it. Theodore Rümpler (1886, 233) proposed that peyote be removed from *Echinocactus* and placed in the new segregate genus *Anhalonium*, thus making the binomial *A. williamsii*, which soon became widely used throughout Europe and the United States. Much earlier, Lemaire (1839, 1–3) had proposed the name *Anhalonium* for another group of spineless cacti, now correctly classified as *Ariocarpus*. *Anhalonium* must be considered a later synonym for *Ariocarpus* unless it has been formally conserved, so according to the *International Code of Botanical Nomenclature* (Greuter et al. 1988, 65), it cannot be validly used as a generic name for any plant. The *Ariocarpus* species superficially resemble peyote because they are spineless geophytes, but they are clearly different genera (Anderson 1963 and 1964).

In 1887, Dr. Louis Lewin received the dried peyote labeled "Muscale Button" from the U.S. firm of Parke, Davis and Company in Detroit (which had obtained the material from Dr. John R. Briggs of Dallas, Texas, earlier that

FIGURE 8.1. The first illustration of peyote appeared in *Curtis's Botanical Magazine* in 1847 (plate 4296).

year) (Bruhn and Holmstedt 1974, 358–60). Lewin used some of this plant material in chemical studies and found numerous new alkaloids (see chapter 7). He also boiled some of the dried "buttons" in water to restore something of their living appearance, and he gave them to the German botanist Paul E. Hennings of the Royal Botanical Museum in Berlin. Hennings noted that Lewin's plant material appeared similar to the plant called "*Anhalonium*" *williamsii* (*Echinocactus williamsii* Lemaire ex Salm-Dyck) but apparently differed somewhat in the form of its vegetative body, namely in the characteristic wool-filled center of the plant. Hennings (1888, 411) decided that the dried plant material given to him by Lewin was that of a new species, which he formally named *Anhalonium lewinii* in honor of his colleague.

Hennings' description was accompanied by two drawings, one of the new species, *A. lewinii*, and the other of the older species, *A. williamsii*. The illustration of *A. lewinii* shows a high mound of wool in the center of the plant. Apparently, the drawing, which had been made from the dried plant material that Lewin or Hennings had boiled in water, was an incorrect reconstruction of what had been the original appearance of the plant. When the top of a peyote dries, the soft fleshy tissue is reduced greatly in volume, whereas the wool does not decrease in size at all. Therefore, the proportion of wool to what formerly was the fleshy or vegetative part is greatly increased in the dried button. This phenomenon presumably caused Hennings and Lewin to believe that they had a new species of peyote, when in actuality, the plant material they had studied was that of "*Anhalonium*" *williamsii*. Bruhn and Holmstedt (1974, 384–85) have concluded that Lewin's plant material known as "*Anhalonium*" *williamsii* was, in fact, the southern species of peyote, *Lophophora diffusa*. The specimens that Hennings described as the new species "*Anhalonium*" *lewinii* belong to the northern species of peyote, *L. williamsii*.

Additional confusion concerning the botanical classification of peyote occurred when the American botanist John Coulter (1891, 129) transferred peyote to *Mammillaria*, a genus commonly called the pincushion or nipple cactus. Then, three years later, Voss (1894) confused things even more by placing peyote in *Ariocarpus*, the valid name for a distinct—and quite different—group of plants that had been called *Anhalonium* also. Finally, Coulter (1894, 131–32) proposed a new genus for peyote alone: *Lophophora*. This helped clarify the nomenclatural situation because peyote had been included in at least five different genera of cacti by the end of the nineteenth century, but the group of plants commonly called and used as peyote is unique within the cactus family and deserves separation as the distinct genus *Lophophora*.

Beginning about 1900, numerous forms and variants of peyote were collected in the field and sent to cactus collectors and horticulturists in Europe and the United States. The highly variable peyote plants, not seen as part of natural populations but as individual specimens in pots, were often described as new and different species. None of the taxonomic studies, however, were based on careful fieldwork, so little was known of the nature and range of variations within naturally occurring peyote populations. By midcentury, greater accessibility of peyote locations in Texas and Mexico permitted exten-

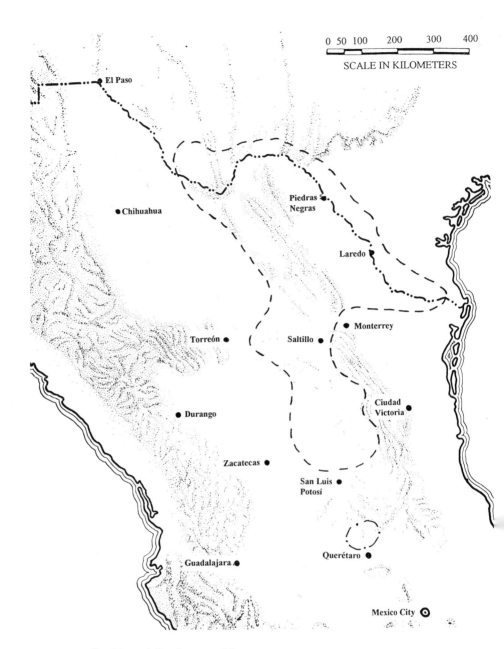

FIGURE 8.2. Natural distribution of the two species of peyote, *Lophophora williamsii* (dashed line) and *L. diffusa* (just north of Querétaro).

sive fieldwork, which has shown that plants of the genus *Lophophora*, especially in the north and central regions of its distribution, are highly variable with regard to vegetative characters (i.e., color, rib number, size, etc.). The number and prominence of ribs, slight variations in color, and the condition of trichomes, or hairs, have tended to be three of the main characters that have delineated many of the proposed species and varieties of peyote; however, these characters vary so greatly even within single populations that they are an insufficient basis for separating species—if a species is considered to be a genetically distinct, self-reproducing natural population.

Field and laboratory studies (Anderson 1969, 299–303) show that there are two major and distinct populations of peyote which represent two species (fig. 8.2). The first, *Lophophora williamsii*, the commonly known peyote cactus that is so widely used by Native Americans, comprises a large northern population, extending from western and southern Texas southward along the high plateau land of northern Mexico. This variable and extensive population reaches its southern limit in the state of San Luis Potosí, where, near the junction of federal highways 57 and 80, for example, it forms large, variable clumps. The second species, *L. diffusa*, is a more southern population that occurs in the dry central area of the state of Querétaro. This species differs from the better-known *L. williamsii* in that it is yellowish-green rather than blue-green in color, lacks any type of ribs or furrows, has poorly developed podaria (elevated humps), and is a softer, more succulent plant.

Common Names of Lophophora

The study of peyote has frequently been confused because the plant received so many different common names. Fr. Bernardino de Sahagún (1938, 3:118) first described the plant in 1560 when he referred to the use of the root "peiotl" by the Chichimeca of Mexico. The two most commonly used names, "peyote" and "peyotl," are modifications of that ancient word, although the actual source and meaning are disputed.

At least three theories have been proposed to explain the etymology of "peiotl." Several scholars have suggested that the term "peyote" came from the Aztec word "pepeyoni" or "pepeyon," which means "to excite" (Schultes 1937). A derived word from this is "peyona-nic," meaning "to stimulate or activate." A similar proposal was made by Reko and extensively discussed by Schultes (1937), who suggested that the term "peyote" came from the different Aztec word "pi-youtli," meaning "a small plant with narcotic action." This somewhat narrow interpretation of the kind of action should perhaps be broadened to mean "medicinal" rather than "narcotic," because Native Americans certainly would have thought of the actions of the plant in the former context.

Probably the most widely accepted etymological explanation for the origin of the term "peyote" was suggested by de Molina (1880), who claimed that it comes from the Nahuatl word "peyutl," which means, in his words, "*capullo de seda, o de gusano*" (80). This phrase, translated from Spanish, means "silk cocoon or caterpillar's cocoon." De Molina's explanation, therefore, proposed that the original word was applied to the plant because of its appearance rather than its physiological action. Certainly one of the most distinctive characteristics of peyote is the numerous tufts of white wool or hair. Dried plant material has an even greater proportion of the "silky material," most of which must be plucked prior to eating.

The presence of these woolly hairs seems to be significant because some other pubescent (hairy or woolly) plants, not even cacti, have occasionally been called peyote by Mexicans. Examples of such non-cacti are *Cotyledon caespitosa* of the family Crassulaceae and *Cacalia cordifolia* and *Senecio hartwegii* of the family Asteraceae (Compositae). These plants have little in common with the peyote cactus except for their pubescence and the fact that sometimes they have been used medicinally. The Mexican Spanish word "piule," which is generally translated to mean "hallucinogenic plant," may have come indirectly from the word "peyote." R. Gordon Wasson (1965, 166–67), who studied many psychotomimetic plants and fungi, suggested that "peyotl" or "peyull" became "peyule," which was further corrupted into "piule." "Piule" is also applied to *Rivea corymbosa* (Convolvulaceae).

Other names that are apparently variations in spelling (and pronunciation) of the basic word "peyote" or "peyotl" include "pejote," "pellote," "peote," "peyori," "peyot," "pezote," and "piotl." The many tribes of Native Americans who use peyote have words for the plant in their own languages, but many also know and use the word "peyote" as well. Some of the tribes and their common names for peyote are

Comanche	*wokowi* or *wohoki*
Cora	*huatari*
Delaware	*biisung*

Huichol	*hicouri, hikuli, hicori, jicori,* and *xicori*
Kickapoo	*pee-yot* (a naturalization of the term "peyote" into their language)
Kiowa	*seni*
Mescalero Apache	*ho*
Navajo	*azee*
Omaha	*makan*
Opata	*pejori*
Otomi	*beyo*
Taos	*walena*
Tarahumara	primarily *híkuli,* but also *híkori, híkoli, jíkuri, jícoli, houanamé, híkuli wanamé, híkuli walúla saelíami,* and *joutouri*
Tepehuane	*kamba* or *kamaba*
Wichita	*nezats*
Winnebago	*hunka*

Numerous other common names applied to *Lophophora* include

biznaga (carrot-like or worthless thing)—commonly applied to many globose cacti
cactus pudding
challote—used principally in Starr County, Texas, one of the major collecting sites for peyote in the United States
devil's root
diabolic root
dry whiskey
dumpling cactus
Indian dope
moon, the "bad seed," "p"—applied to peyote by drug users in the United States
raíz diabólica (devil's root)
tuna de tierra (earth cactus)
turnip cactus
white mule

Part of the confusion regarding the numerous common names for *Lophophora* results from their frequent use for cacti of other genera or plants from another family entirely.

Plants Confused with Peyote

Two factors have led to the confusion of various plants being called "peyote": (1) a similarity of appearance because of pubescence, a globose shape, or growth habit, and (2) a similar physiological effect or use for medicinal or religious purposes. In fact, most of the plants that are sometimes called "peyote" possess both characters.

Many alkaloid-containing cacti are commonly called "peyote," but they are not in the genus *Lophophora*, and even though some of the alkaloids are the same, they probably have few or no physiological actions similar to the true peyote. In the following list are cacti that have at one time or another been called "peyote," or the Spanish diminutive "peyotillo."

> *Ariocarpus agavoides:* A very small member of this genus found only in Tamaulipas. It is also called "magueyitos" (little agaves) by local inhabitants, and children sometimes collect and eat the plants for their sweetish taste.
>
> *A. fissuratus:* More frequently called "living rock" or "chautle" but also "peyote cimarrón" (wild peyote). It is called "sunamí" by the Tarahumaras, who consider it more potent than peyote, although it does not contain mescaline.
>
> *A. kotschoubeyanus:* Usually called "pezuña de venado" (cloved hoof of a deer) or "pata de venado" (deer's foot).
>
> *A. retusus:* Usually called "chautle" or "chaute." It is known as "false peyote" to the Huichols because it has undesirable side effects.
>
> *Astrophytum asterias:* Surprisingly similar in appearance to *Lophophora* and sometimes cultivated with peyote in "peyote gardens" in south Texas.
>
> *A. capricorne:* Also called "biznaga de estropajo" (carrot-like vegetable sponge).
>
> *A. myriostigma:* Called "peyote cimarrón" (wild peyote), "mitra" (miter), and "birrete de obispo" (bishop's cap or miter).
>
> *Aztekium ritteri:* Another small, globose cactus with superficial resemblance to *Lophophora*.
>
> *Mammillaria longimamma:* Sometimes called "peotillo."
>
> *M. pectinifera:* Known locally in Puebla as "cochinito" (little pig).
>
> *Obregonia denegrii:* Also called "obregonita."
>
> *Pelecyphora aselliformis:* Commonly called "peyotillo" and sold as such

in the native markets. It contains some of the alkaloids possessed by *Lophophora*, including small amounts of mescaline.

Strombocactus disciformis: Similar in appearance to *Lophophora* and occurring in the same general area as *L. diffusa*.

Turbinicarpus pseudopectinatus: Small, globose, and whitish, though the color is due to spines rather than wool.

Other plant families, including the Asteraceae (Compositae), Crassulaceae, Fabaceae (Leguminosae), and Solanaceae, also have representatives that occasionally are called "peyote." A member of the Asteraceae was first described as a type of peyote by the Spanish physician Francisco Hernández (1790, 70–71) in his early study of the plants of New Spain. He described two peyotes: the first, *Peyotl Zacatecensi*, clearly was *Lophophora*, whereas the other, *Peyotl Xochimilcensi*, apparently was *Cacalia cordifolia*, of the Asteraceae, which had "velvety tubers" and was used medicinally. Other sunflowers of the closely related genus *Senecio* have also been called such things as "peyote del Valle de México" and "peyote de Tepic."

Although "mescal" is the correct name for the alcoholic beverage obtained from the century plant, *Agave angustifolia* and several other species, the word was also used by missionaries and officials of the Bureau of Indian Affairs for peyote. Possibly this was an attempt to confuse Congress and the public into thinking that peyote was an "intoxicant" similar to alcohol, so that it could be prohibited under anti-alcohol legislation, but it may just have been a case of incorrect information perpetuated unwittingly.

The name "mescal bean" has also been applied incorrectly to peyote, but actually it is the common name of *Sophora secundiflora* of the Fabaceae (Leguminosae). The beans of this plant contain cytisine, a toxic pyridine that causes nausea, convulsions, hallucinations, and even death if taken in too-large quantities (Schultes and Hofmann 1979, 74–75). The colorful red beans have been used by Native Americans for centuries in Mexico and the United States for medicinal and ceremonial purposes, and sometimes the seeds of this desert shrub are worn as necklaces by the leaders of peyote ceremonies. The stimulatory and hallucinatory nature of these beans probably led to its confusion with peyote, especially because the latter occasionally was called "mescal." The probable relationship of the old mescal bean ceremony and the modern peyote religion (discussed in chapter 2) also may have led to confusion.

Peyote has also been referred to as the "sacred mushroom," probably because of the similar appearance of dried peyote tops and dried mushrooms. Also, some mushrooms can produce color hallucinations similar to those of peyote. The Spaniards first misidentified peyote as a mushroom late in the sixteenth century when they stated that the Aztec substance "teonanacatl" and peyote were the same, a mistake perpetuated by the American botanist William E. Safford (1915, 311). He and other reputable scientists insisted that there was no such thing as the sacred mushroom "teonanacatl"; they believed that it was simply the dried form of peyote. The problem was resolved when psychotomimetic mushrooms were rediscovered in 1936 and definitely linked to early Mexican ceremonies. In recent years at least fourteen species of psychoactive mushrooms have been identified in the genera *Psilocybe, Stropharia, Panaeolus,* and *Conocybe* of the family Agaricaceae. It is evident that they are well known to Native Americans in Mexico (Schultes and Hofmann 1979, 144–53).

Another plant that has occasionally been confused with peyote is "ololiuhqui," which is now known to be *Rivea corymbosa* of the Convolvulaceae. Ololiuhqui has been widely used by Native Americans in the Sierra Madre Occidental of Mexico as an aphrodisiac, a cure for syphilis, an analgesic, a cure for colds, a stimulating tonic, a carminative (to relieve colic), a help for sprains and fractures, and for relief of pelvic cramps in women (Schultes 1937, 74). Recent studies have shown that the plant contains several potent chemicals that are ergot alkaloids closely related to LSD (Schultes and Hofmann 1979, 159–63). The effects are somewhat similar to those of peyote: stimulatory at first and later producing color hallucinations. Native Americans could easily see many "divine" actions resulting from ingestion of *Rivea* seeds, and it is not difficult to understand why they and others may have confused it with peyote, another "divine" plant.

The following summary of the Mexican plants and fungi that, like *Lophophora*, are psychotomimetic gives the ancient Mexican name, the botanical name or names, the plant or fungus family, and one or more of the main psychoactive substances (Wasson 1965, 164–75).

> Picietl = *Nicotiana rustica* (Solanaceae): A species of tobacco containing exceptionally high quantities of nicotine.
> Teonanacatl = *Psilocybe* spp.
> *Panaeolus campanulatus* var. *sphinctrinus*

Stropharia cubensis
Conocybe spp.

All of the above are in the fungus family Agaricaceae. The psychoactive substances are psilocybin and psilocin.

Pipiltzintzinli = *Salvia divinorum* (Lamiaceae): The psychoactive principle of this plant is as yet undetermined.
Ololiuhqui = *Rivea corymbosa* (Convolvulaceae)
Tlitlitzen = *Ipomoea violacea* (Convolvulaceae)

Both of the above members of the Convolvulaceae contain the ergot or lysergic acid alkaloids; LSD is a synthetic derivative, but does not occur naturally.

Marijuana is one of the best-known and most widely used substances currently classified by many as a psychotomimetic. Others, however, seriously question whether it actually is a hallucination-producing plant (at least in the way that it is used by most people)—and it is of Old World origin. Marijuana, which is obtained from the genus *Cannabis* of the angiosperm family Cannabaceae (Schultes et al. 1974, 360–62), is psychoactive but has quite different effects than does peyote (see chapter 6).

Morphology

Morphological studies (Boke and Anderson 1970) have provided much information about the evolution and relationships of the cacti. Investigations of both vegetative and reproductive parts support the proposal that *Lophophora* is a distinct genus consisting of two species.

Vegetative Parts

The growing point, or apical meristem, located in the depressed center of the plant, is relatively large and similar to those found in other small cacti. The young leaf, which arises from the meristem, is difficult to distinguish from the expanding leaf base and subtending axillary bud. The leaf base, usually separated from the actual leaf by a slight constriction, grows rapidly to become the podarium, rib, or tubercle. Thus the leaf base functions as the photosynthetic or food-producing part of peyote. The vestigial leaves of seedlings are often large enough to be identified with sufficient magnification, but they are never more than a microscopic hump in the vegetative shoot of mature peyote plants.

Spines occur only on young seedlings; adult plants produce spine primordia, but they rarely develop into spines.

The caespitose, or several-headed, condition of the peyote cactus apparently occurs through the activation of adventitious buds that appear on the tuberous part of the root-stem axis below the crown. Such growth often is the result of injury and almost always occurs if the top of the plant is cut off. However, some populations of peyote seem to have a greater tendency to develop the caespitose condition than do others.

Epidermal cells, usually five- to six-sided and papillose (nipple-like), have cell walls only slightly thicker than those of the underlying parenchyma cells. Sometimes a hypodermal layer can be recognized early in development, but as the stem matures, it does not become specialized and never differentiates from the underlying palisade tissue. Normally the epidermis is covered by both cuticle and wax; the latter substance is primarily responsible for the blue-green or glaucous coloration of *L. williamsii*. Stomata are abundant, especially on the younger, photosynthetically active parts of the vegetative body. They are paracytic and usually subtended by large intercellular spaces. The subsidiary cells of a stoma usually are about twice the size of neighboring epidermal cells. Trichomes are persistent for many years in the form of tufts of hairs or "wool" arising from each areole. They tend to be uniseriate on the younger areoles but are often multiseriate on older ones.

Ergastic substances are evident in the cortex of peyote. Usually they are druses of calcium oxalate that often exceed 250 micrometers in diameter, but which rarely are found within one millimeter of the epidermal layer. These anisotropic crystals can be easily seen if fresh or paraffin-embedded sections are examined in polarized light. Mucilage cells do not occur in the vegetative parts of peyote but are found in flowers and young fruits.

The chromosome number of peyote, like most other cacti, is $2n = 22$. The root-tip chromosomes are quite small, and apparently there is no variation from the basic chromosome number of the Cactaceae, which is $n = 11$.

Reproductive Parts

Peyote flowers, in contrast to those of other cactus genera such as *Echinocactus* and most of the genus *Thelocactus*, have naked ovaries, or the absence of scales on the ovary wall—a character shared with the flowers of *Mammillaria*, *Ariocarpus*, *Obregonia*, and *Pelecyphora*. In *Lophophora*, all floral parts are borne on the perianth tube above the ovule-containing cavity. The flower color of *Lophophora* varies from deep reddish-pink to nearly pure white; those of *L. diffusa* rarely exhibit any red pigmentation, making them usually appear white or sometimes a light yellow because of the reflection of yellow pollen from the center of the flower. Development of peyote flowers is much like that of *Mammillaria*.

Pollen of *Lophophora* is highly variable (Anderson and Stone 1971; Leuenberger 1976, 112, 142, 193, 291). Pollen of the Dicotyledonae tend to have three apertures or pores, while those of the Monocotyledonae usually have only one aperture. Peyote pollen varies greatly in aperture number, the northern population having 0 to 18 and the southern population 0 to 6. The grains are spheroidal, polyporate, with reticulate exine bearing small spicules, and 40 to 53 micrometers in diameter. The varying numbers of colpae or apertures produce about twelve different geometric shapes. Such a variety from a single species or even population is rare in flowering plants.

The pollen of *L. diffusa* has less variation than that of *L. williamsii*, and it also has a much higher percentage of grains that are of the basic tricolpate (three-aperturate) type. Thus the basic dicotyledon pattern is best observed in the southern population, whereas more-complex grains occur in the northern localities. Small tricolpate grains probably are more typical of the ancestors of the cacti, and the more elaborate geometric designs of *L. williamsii* seem to represent greater evolutionary divergence and specialization.

Fruits of peyote develop and remain hidden in the apical wool for about a year; then they elongate rapidly at maturity to protrude above the woolly center of the plant. The fruits of *Lophophora* are similar to those of *Obregonia* in that usually only the upper half contains seeds, whereas *Ariocarpus* fruits are completely filled with seeds.

The seeds of *Lophophora* are black and verrucose (warty), with a large, flattened, "basal," whitish hilum. They are virtually identical to those of *Ariocarpus* and *Obregonia*, although there are some minor structural differences of the testa.

Lophophora seems to stand by itself in possessing a particular combination of morphological characters unlike any other group of cacti. Its nearest relatives appear to be the genera *Echinocactus*, *Obregonia*, *Pelecyphora*, *Ariocarpus*, and *Thelocactus*. The character of seeds, seedlings, areoles, and fruits certainly supports the contention that peyote belongs in the subfamily Cactoideae, tribe Cacteae, subtribe Echinocactinae (IOS Working Party 1990).

Zimmerman (1985, 161), in a detailed cladistic analysis of several cactus genera, showed that *Lophophora* is a distinct line from the most closely allied genera, *Aztekium* and *Strombocactus*. Likewise, Leuenberger (1976, 193) reported that pollen data show a relationship to *Aztekium* and *Strombocactus*, as well as *Turbinicarpus*. Robert Wallace of Iowa State University, on the other hand, reports that preliminary DNA studies of peyote indicate that it is most closely related to *Obregonia* (pers. comm.).

BIOGEOGRAPHY

The genus *Lophophora* is one of the most wide-ranging of all the plants occurring in the warmer regions of North America. It occurs throughout the Chihuahuan Desert and in the Tamaulipan Thorn Forest. It has a latitudinal distribution of about 1,300 kilometers (800 miles), from 20°54' to 29°47' north latitude (see fig. 8.2). Within the United States, *L. williamsii* is found in the Rio Grande region of Texas, with small populations occurring in western Texas in Presidio and Val Verde Counties. It has also been reported from the Big Bend region in Brewster County, but its main distribution in Texas is in the Rio Grande valley eastward from Laredo.

Peyote extends from the international boundary southward into Mexico in the basin regions between the Sierra Madre Occidental and the Sierra Madre Oriental to Saltillo, Coahuila; this vast expanse of Chihuahuan Desert in northern Mexico covers about 150,000 square kilometers (60,000 square miles). Just south of Saltillo the range of peyote narrows, is interrupted by mountains, and then expands again eastward into the foothills of the Sierra Madre Oriental and westward into the state of Zacatecas. It extends southward nearly to the city of San Luis Potosí, where its distribution terminates (figs. 8.3 and 8.4). The southern population of peyote, that of *L. diffusa*, is restricted to a high desert region in the state of Querétaro. This area of about 775 square kilometers (300 square miles) is isolated from the large northern populations by high, rugged mountains (figs. 8.5 and 8.6).

I have observed and collected peyote throughout its known range, except for the Big Bend region of west Texas. There was a good population of *Lophophora* near Shafter some years ago, but now the area is fenced and posted, so one cannot visit the population. I have searched extensively in an area on the west side of Chilochotal Mountain, Big Bend National Park, where Barton

FIGURE 8.3. The habitat of *Lophophora williamsii*. This photo was taken one kilometer northeast of the junction of Highways 57 and 80 near El Huizache in the state of San Luis Potosí. Vegetation includes *Yucca filifera* (yucca), *Prosopis glandulosa* (mesquite), *Agave lechuguilla*, *Opuntia engelmannii* (prickly pear cactus), and *Myrtillocactus geometrizans*. Peyote is growing at the base of the creosote bush (*Larrea tridentata*) in the right center of the picture.

Warnock collected peyote many years ago. No one has seen it there for several years, but Tom Alex, an archaeologist and ranger at the park, believes that this peyote population at the western edge Chilochotal Mountain is (or was) a natural one (pers. comm.). The most recent report of anyone seeing peyote in the park was on the east side of Chilochotal Mountain about twenty-five years ago. Native Americans occupied several sites at the base of the mountain for at least four thousand years, which was determined by midden and other data. There is definite evidence that the Tarahumaras came into the area. Alex believes it is possible that Native Americans, perhaps with the aid of Anglos during the 1960s, collected the population to extinction, or nearly

FIGURE 8.4. Clump of *Lophophora williamsii* in the San Luis Potosí habitat shown in figure 8.3.

so. There are additional reports of peyote occurring at other sites in Presidio County, as well as in Val Verde County just west of the Pecos River (Michael Powell, pers. comm.).

Three factors apparently are responsible for the discontinuous distribution of *Lophophora* between the large northern and the smaller southern population in Mexico: (1) extensive saline flats in the Rio Verde region east of the city of San Luis Potosí, (2) the formidable Sierra Gorda Mountains, which are an extension of the Sierra Madre Oriental, and (3) high elevations, even in the broad valleys. The relatively high desert area in Querétaro apparently is an isolated pocket of the Chihuahuan Desert.

There are great elevation differences from the north to the south within the total range of peyote; the Rio Grande peyote occurs at an elevation of about 50 meters (150 feet), but in the southern portion of its range in the state of San Luis Potosí, peyote is found at nearly 1,850 meters (6,000 feet) elevation. The elevation of the southern population in Querétaro is about 1,500 meters (5,000 feet) (Anderson 1969, 302).

FIGURE 8.5. The habitat of *Lophophora diffusa*. This photo was taken 45 kilometers north of Cadereyta des Montes and 18 kilometers north of Vizarrón on Highway 20 in Querétaro. Vegetation is mostly *Fouquieria formosa* (ocotillo) and *Larrea tridentata* (creosote bush). *Lophophora diffusa* grows only in a limited area on either side of a 15- to 20-kilometer stretch of highway in Querétaro.

ECOLOGY

Peyote occurs primarily in the Chihuahuan Desert, which is a type of warm-temperate desert biome. This region has considerable variation in both topography and vegetation, which has prompted ecologists to describe numerous subdivisions. Unfortunately, these subdivisions are not alike nor have they received the same names. Following the classification of Mexican botanist Jerzy Rzedowski (1978, 237–61), peyote occurs within the vast region called the Matorral Xerófilo. This area has been subdivided in various ways, even by Rzedowski himself (1966, 219–20). He described the Chihuahuan Desert as consisting of (1) the microphyllous desert scrub, which has shrubs that are leafless or have small leaves and are represented by such plants as *Larrea tridentata*, *Prosopis glandulosa*, and *Flourensia cernua*; and (2) the "rosettophyl-

Ecology : 171

FIGURE 8.6. Cluster of *Lophophora diffusa* growing in the Querétaro habitat shown in figure 8.5.

lous" desert scrub, with many plants bearing rosettes of leaves, such as *Agave lechuguilla, Hechtia* spp., and *Yucca* spp. Probably neither of these vegetation subdivisions can be considered climax communities, nor even formations, because there is continuous mixing of the two life forms. Since there is such confusion between these two subdivisions, perhaps Cornelius H. Muller's (1947, 38) general term "Chihuahuan Desert Shrub" is one of the best with which to describe the general area in which peyote occurs.

The well-isolated southern population apparently is outside the region normally included within the Chihuahuan Desert. However, the presence of *Larrea tridentata* and other plants typical of this type of desert is an indication that it should, indeed, be included within the Chihuahuan Desert.

The soils of the Chihuahuan Desert Shrub are limestone in origin and have a basic pH, from 7.9 to 8.3. These soils can also be characterized as having more than 150 ppm (parts per million) calcium, at least 6 ppm magnesium, strong carbonates, and no more than trace amounts of ammonia. The soils test negatively for iron, chlorine, sulfates, manganese, and aluminum. Phosphorus and potassium vary somewhat throughout the range, but in most localities occur in trace amounts or are not present at all. Soils from the southern locality in Querétaro are not different from those to the north (Anderson 1969, 302).

As stated earlier, peyote occurs in diverse habitats of the Chihuahuan Desert and into the Tamaulipan Thorn Forest, and no particular plants are associated with it in all localities. Plants commonly found with peyote in the Chihuahuan Desert are

Acacia spp. (acacia)
Agave lechuguilla (lechuguilla)
A. striata
A. stricta
Ariocarpus retusus
Astrophytum myriostigma (Bishop's cap cactus)
Condalia lycioides (lotebush)
Coryphantha spp.
Dasylirion wheeleri (sotol)
Echinocactus horizonthalonius (eagle claws cactus)
E. platyacanthus
Echinocereus pectinatus (rainbow cactus)
E. reichenbachii
Euphorbia antisyphylitica (waxplant)
Ferocactus echidnae
F. hamatacanthus
F. pilosus
Flourensia cernua (tarbush)
Fouquieria splendens (ocotillo)
Hechtia glomerata
Jatropha dioica (leatherplant)
Koeberlinia spinosa (crucifixion thorn)
Larrea tridentata (creosote bush)
Leuchtenbergia principis (Agave cactus)
Mammillaria spp. (fishhook or nipple cactus)
Mimosa spp. (cat claw)
Neolloydia conoidea

Opuntia engelmannii
O. leptocaulis (pencil cholla)
O. stenopetala
O. tunicata (cholla)
Parthenium incanum (guayule)
Prosopis glandulosa (mesquite)
Yucca carnerosana
Y. filifera (yucca)

The northeastern part of the geographical range of peyote is outside the Chihuahuan Desert, located in a region referred to by Correll and Johnston (1970, 7–9) as the Rio Grande Plains. It has also been called the South Texas Plains and Tamaulipan Brushlands, for it extends southward from the international border into the Mexican state of Tamaulipas. In contrast to the more open Chihuahuan Desert vegetation, the Tamaulipan Brushlands or Thorn Forest consists of dense, thorny trees and shrubs. Rzedowski (1978, 213–14) refers to this region as a variant of the "Mezquital," which he places in the larger category "Bosque Espinoso." It has also been referred to as the "Tamaulipeca" region of the "Matorral Xerófilo" (Vite-González et al. 1990). Characteristic vegetation of this region includes

Acacia farnesiana (huisache)
A. greggii (cat claw)
A. rigidula (black brush)
Agave lechuguilla (lechuguilla)
Bursera fagaroides
Celtis pallida (granjeno)
Cercidium macrum (retama)
Cordia boissieri (anacahuite)
Echinocactus texensis (horse crippler)
Karwinskia humboldtiana (coyotillo)
Leucophyllum spp. (cenizo)
Mammillaria spp. (nipple cactus)
Opuntia engelmannii (prickly-pear cactus)
O. leptocaulis (pencil cholla)
Porlieria angustifolia (guayacan)
Prosopis glandulosa (mesquite)
Thelocactus setispinus

Yucca elephantipes
Y. filifera (yucca)
Ziziphus obtusifolia (clepe)

Of course not all perennial plants growing with peyote have been cited, but this information indicates that peyote occurs over a broad range of vegetation types within the Chihuahuan Desert and Rio Grande Plains.

The climatic data from the regions in which peyote grows have been analyzed to obtain an "index of aridity." Using the index of aridity devised by Soto Mora and Ernesto Jáuregui O. (1965, 26–28), peyote is found to tolerate a very wide range of climatic conditions: precipitation from 175.5 millimeters up to 556.9 millimeters per year, maximum temperatures from 29.1°C to 40.2°C, and minimum temperatures from 1.9°C to 10.2°C. There is also a variation in the time of year that precipitation occurs. Rains typically fall in the late spring and summer in the Chihuahuan Desert and Rio Grande Plains, but in certain areas some winter rains do fall. There are peyote populations in both types of areas, so probably they should be classified as being in intermediate, rather than strictly summer, rainfall regions. The modified index of aridity, based on the relationship of temperatures and precipitation, shows that *Lophophora* exhibits a wide range of aridity, between 64 and 394. It also appears that the index of aridity is related to elevation, although there are some definite exceptions, such as in Querétaro, where there is a relatively high elevation (about 1,500 meters or 5,000 feet) but an index of aridity that is over 115. This southern habitat, although of high elevation, may be especially arid because of the proximity of surrounding high mountains that cause a more intensified rain shadow.

CHARACTERISTICS

Peyote consists of populations that are not only wide-ranging geographically, but which are also variable in topographical locations, appearance, and methods of reproduction. Peyote is commonly found growing under shrubs such as *Prosopis glandulosa* (mesquite), *Larrea tridentata* (creosote bush), and the rosette-leafed plants, such as *Agave lechuguilla* and *Hechtia glomerata*; at other times, however, it grows in the open with no protection or shade of any kind. In some areas, such as in the state of San Luis Potosí, peyote sometimes grows in silty mud flats that become temporary, shallow freshwater lakes during the rainy season. Peyote has even been found growing in crevices on steep limestone cliffs in west Texas.

The appearance of peyote also varies widely, especially in the species *L. williamsii*. In some cases the plants occur as single-headed individuals, and in others they become caespitose, forming dense clumps up to two meters across with scores of heads. Plants in Texas do not seem to form large clumps as often as those in the state of San Luis Potosí, but plants with several tops can arise as the result of injury by grazing animals or other factors. Caespitose individuals are also produced by harvesting the tops. In Texas, for example, collectors normally cut off the top of the plant, leaving the long, carrot-shaped root in the ground; the subterranean portion soon calluses and in a few months produces several new tops to replace the one that was cut off.

The number of ribs in a single head varies widely. Rib number and arrangement appear to be factors of age as well as responses to the environment. Rib number within a single, genetically identical clone may vary from four or five in very young tops, up to fourteen in large, mature heads (see fig. 8.4). At other times there are bulging podaria instead of distinct ribs. Field studies have shown that rib number and variation apparently are due to localized interactions between genotype and environment. Because of the high degree of variation occurring in a single population, rib characteristics alone are of little value in the delimitation of formal botanical taxa.

Reproduction occurs mainly by sexual means. The plants flower in the late spring and throughout much of the summer, and the ovules, which are fertilized during that season, mature into seeds a year later. The fruit, which arises from the center of the plant late in the spring or early in the summer, rapidly elongates into a pink or reddish cylindrical structure up to about 1.5 centimeters (0.5 inch) in length. Within a few weeks these fruits mature: their walls dry, become paper thin, and turn brownish. Later in the summer, usually as a result of wind, rain, or some other climatic factor, the fruit wall ruptures and the many small black seeds are released. The heavy summer rains then wash the seeds out of the sunken center of the plant and disperse them.

Another method of reproduction in peyote is by vegetative or asexual means. Many plants produce "pups," or lateral shoots, which arise from lateral areoles. These new shoots can often root and survive when broken off if they have attained sufficient size. If these new portions successfully grow into new plants, they are genetically identical to their parents. Surprisingly, peyote plants rarely rot if injured or cut, so excised pieces will readily form adventitious roots and can become independent plants.

Evolution

The evolutionary history of the cacti is not documented by fossils because their succulent vegetative parts did not lead to preservation as fossils in the arid climate of the desert. The highly specialized cactus has few characteristics of its distant ancestors, but it does appear that the tropical leafy cactus *Pereskia* may represent a form that has changed little from the non-cactus ancestral types. It and many of the more specialized cacti have several characteristics similar to the other eleven families of the order Caryophyllales, in which the cacti are often placed. Most of these families, for example, have a glandular anther tapetum, simultaneous divisions of the microspore mother cells, trinucleate pollen grains, a curved embryo with a massive nucellus, the presence of perisperm rather than endosperm, either basal or free-central placentation, betalain pigments rather than the usual anthocyanins, anomalous secondary thickening of the xylem walls, a distinctive type of sieve tube plastid, and succulence (Cronquist 1981, 235–76).

Clearly, the Cactaceae is monophyletic, as are each of the three subfamilies: Pereskioideae, Opuntioideae, and Cactoideae. All of these subfamilies have undergone independent evolution since arising from a common ancestor. *Pereskia*, one of two genera within the Pereskioideae, shows what many believe are "primitive" features probably possessed by the ancestor of the cacti. Apparently the living representatives of the cacti are terminal points of a highly branched evolutionary history, and ancestors no longer exist. Therefore, we must work with characters of living representatives to draw any conclusions regarding the past evolutionary history of the cacti, a procedure of speculation at best.

Lophophora is a member of the subfamily Cactoideae, which contains about 85 percent of the cactus species and has a great diversity of habit, floral structures, and vegetative characters. The group has also undergone parallel evolution in North and South America. Some species of the South American genera *Frailea* and *Gymnocalycium* have great morphological similarity to *Lophophora*. DNA studies indicate that the Cactoideae is clearly a monophyletic group (R. Wallace, pers. comm.).

Peyote is a highly derived member of the Cactoideae, with its closest relative probably being *Obregonia* (R. Wallace, pers. comm.). Zimmerman (1985, 160–61) also noted that *Lophophora* is closely related to *Aztekium* and *Strombocactus*, differing morphologically only by seed characters. He also showed in a cladogram that *Ariocarpus* is within this group of several closely related genera.

Certain evolutionary trends appear evident in the two species of *Lophophora*. Pollen of *L. diffusa*, because of its higher percentage of the basic tricolpate type of grain, could be considered more primitive than that of *L. williamsii*. Likewise, Todd (1969) and Bruhn and Holmstedt (1974) have shown that certain of the more elaborate alkaloids are either absent or in lesser amounts in *L. diffusa*. This, they feel, indicates that *L. diffusa* may not have evolved and diversified to as great an extent chemically as has *L. williamsii*. Also, the greater variation of the vegetative body of *L. williamsii*, and its more varied habitats and a wider distribution, probably show a more diverse and highly evolved gene pool.

Lophophora probably arose from a now-extinct ancestor that occurred in semidesert conditions in central or southern Mexico. Morphological and chemical diversity may have then appeared in various populations as they slowly migrated northward into drier regions being created by the slow uplift of mountains. Perhaps *L. diffusa* represents one of the earlier forms that became isolated in Querétaro, whereas *L. williamsii* spread more extensively to the northward, producing new combinations of genes that eventually led to a distinct but highly variable species having somewhat different pollen, vegetative characters, and alkaloids from the peyote populations to the south.

CULTIVATION

Most states, as well as the federal government, now prohibit the possession of peyote (see chapter 9). One is in violation of the law even if peyote is grown as part of a horticultural collection, for which there have been arrests and convictions. It is legal, however, to grow peyote in many other countries.

Peyote is easily cultivated and is free-flowering. On the other hand, it takes great patience to grow peyote from seed, as it may take up to five years to obtain a plant that is 15 to 20 millimeters (one-half to three-quarters inch) in diameter. At any stage, however, peyote can be readily grafted onto faster-growing rootstocks such as *Myrtillocactus geometrizans* or *Pereskiopsis velu-*

tina (Cattabriga 1994, 130); this usually triples or quadruples the plant's rate of growth. Japanese nurserymen have obtained peyote plants large enough to flower within a period of 12 to 18 months by grafting the young seedlings onto older, more robust root stocks.

To insure production of fertile seed, it is advisable to out-cross peyote plants by transferring with forceps some stamens containing pollen from the flower of one plant to the stigma of the flower of another. Some cultivators simply transfer pollen with a small brush, but care must be taken to clean the brush after each procedure.

Propagation can also be accomplished by removing small lateral tops from caespitose individuals. The cut "button," or top, should be allowed to callus for a week or two and then planted in moist sand or a mixture of sand and vermiculite. It is wise to dip the freshly cut portion in sulfur or a fungicide powder to avoid infection and facilitate healing. Rooting is best done in late spring or early summer. Eventually a new root system will develop from the top, and the root system from which the top was cut may produce several new heads to make a caespitose individual.

Soil conditions for the cultivation of peyote are not too critical. The natural soil for peyote is of limestone having a basic pH, so one should provide adequate calcium (lime is a satisfactory additive), insure that the soil is slightly basic, and provide good drainage. Peyote should be watered frequently (every four to seven days) in the summer, but very little or none at all in the winter. Fertilizer should be applied while the plants are being watered during the growing season, especially May through July.

Peyote hosts few insect pests and does not need to be treated differently from other cultivated cacti and succulents with regard to pesticides. Greenhouse-grown peyote plants sometimes develop a corky condition; this brownish layer often covers most of the plant and is not natural. Its cause is not known.

The propagation of seeds is a rewarding experience but requires great patience. Seeds should be sowed on fine washed sand or a mixture of sand and vermiculite. Some growers prefer to cover the seeds with 1 to 2 millimeters (about one-eighth inch) of very fine sand, whereas others leave the seeds uncovered, carefully pushing the new seedlings into the sand with a toothpick once the seeds have germinated (Miller 1988, 38). The flat or pot should be covered with a plastic bag or plate of glass with a Grolux or similar lamp placed above. It is important that the light wavelengths be in the range of 400 to 700 nanometers, known as the Photosynthetic Active Radiations (PAR). Otherwise, etiolation and other problems may occur (Cattabriga 1994, 122). The sand should be kept moist to insure high humidity and prevent young plants from drying out as they sprout. Germination usually occurs within two or three weeks, but seedling growth is exceedingly slow. The plants should be transplanted and thinned after they are about 1 centimeter (one-quarter inch) in diameter.

CONSERVATION

It is unclear to what extent humans have affected the distribution of peyote. In some areas large quantities of the plant have been collected, such as in Starr, Jim Hogg, and Webb Counties, Texas; near Matehuala, San Luis Potosí; and in the dry desert valley area of Querétaro. In 1961 I collected *L. diffusa* in a region near the road going north from Vizarrón, Querétaro. In 1967 I returned to the same area but could find no peyote. Farmers living nearby told us that about a year earlier a man from a nearby village, whom they called a "Padre," hired workers to collect all of the peyote that they could find in the region. The farmers did not know why the man had wanted so many plants or what he planned to do with them, but I doubt that they were used for religious or medicinal purposes. Probably they were sold to cactus collectors—or perhaps even destroyed. More recent reports indicate that illegal collecting has continued in Querétaro, and the size of several populations of *L. diffusa* has been greatly reduced. Fortunately, peyote is a common, widespread plant in many areas that are almost inaccessible. We may, however, see considerable disturbance and loss of peyote populations in areas easily reached by man.

Morgan and Stewart (1984, 291–95) noted the decrease in plants in south Texas, and I visited this region to observe peyote in the field as well as to talk to peyoteros and others interested in the plant. About half of the peyote plants in a population in Starr County had been previously harvested, which could be easily determined by carefully digging around clusters of young tops to see the nature of the roots beneath the ground. Part of the shortage in south Texas is simply the ever-increasing demand for peyote buttons by members of the Native American Church, but some who visit the peyote gardens of south Texas do not know how to correctly harvest the tops without damaging the root system.

Two other significant factors have led to a shortage of peyote in Starr, Jim

Hogg, Zapata, and Webb Counties. The first is the locking up of private land, which prevents those who wish to harvest peyote from doing so. Landowners were bothered in the 1960s by "hippies" or members of the drug culture who came to south Texas and illegally entered ranches to collect peyote and camp there. These people are no longer a problem, but Native Americans often come to south Texas to visit the peyote gardens and collect tops for their personal use. In doing this, they often trespass illegally. Landowners have also found that hunting can be a profitable use of the thorn forest, so many have erected high fences to control the movement of game. Groups then lease the right to hunt on the land during the season. The high fences, as well as the danger of being shot during hunting season, have made it difficult to harvest peyote. Some ranchers are willing to lease their land to peyoteros for a period of several weeks so that they may harvest peyote; however, the cost of such leases is quite high, often amounting to several thousand dollars.

A second major problem in south Texas is the destruction of natural vegetation by root plows to facilitate the growing of grass for grazing (Anderson 1995). Root plows destroy virtually all the natural plants except for *Opuntia engelmannii*, which frequently develops into dense thickets on the plowed land. Peyote does not survive.

The long-term future of peyote in south Texas is of considerable concern. Possibly Hispanics and Native Americans who live in the area could be encouraged to salvage peyote from land about to be root plowed. These plants could then be placed into cultivation. However, ranchers would have to be willing to allow peyoteros and others onto their property prior to the plowing, and land for the cultivation of the salvaged plants would have to be found. Clearly, cooperation among landowners, peyoteros, and Native Americans is the only hope for the perpetuation of large stands of peyote in south Texas.

Both the United States and Mexico have laws prohibiting the collection or possession of peyote, with the exception, of course, of licensed peyoteros in the United States. Illegal collecting, however, continues in both countries. In 1991 a shipment of 7,200 living peyote plants of varying ages, including 1,206 of the much more rare *Lophophora diffusa*, was confiscated in The Netherlands (N. P. Taylor, pers. comm.). Interestingly, the shipment was bound from Japan to a Dutch grower who no longer had enough stock plants. He therefore contracted with someone to collect wild plants in Mexico and ship them via Japan to The Netherlands. Recently a man in New Jersey was convicted of possessing peyote in his cactus collection and received a prison sentence.

It is unfortunate that law-enforcement officers and legislators have overreacted to the dangers of peyote, and that those who wish to have one or a few plants in their cactus collections cannot legally do so. Such ill-informed reactions have been occurring for more than a hundred years (see chapter 9). Whether education will reduce the concern of those wanting to outlaw peyote as a "dangerous drug" remains to be seen.

The botany of peyote is now well understood, and current studies will soon determine its relationship to the other cacti. But the future of many natural populations of peyote is threatened by an ever-increasing pressure by humans to collect the plant for religious purposes, and by the destruction of various sites for development and agriculture. Fortunately, the peyote cactus extends over a wide range of North America, and there is little danger that all populations will be destroyed.

CHAPTER NINE

Legal Aspects of Peyote

Almost as soon as European immigrants learned of peyote and its effects on human sensory perceptions, many people felt it would be necessary to control its use. Religious leaders believed it was a device of the devil to keep the indigenous people from adopting Christianity. Some government officials thought it degraded Native Americans, and a few physicians feared it was dangerous to the body and might be used in place of Western medicines. Thus for more than 350 years, there have been efforts to regulate the use of peyote in North America.

Various laws have been created to prohibit its use in the United States and Mexico. Some have assailed these laws, claiming that they prohibit the use of a harmless plant and interfere with the bona fide religious use of peyote as a sacrament. Others have asserted that laws should be more strict to prevent the use of such a "dangerous narcotic" for any purpose whatsoever. And because most lawmakers and citizens have no knowledge of peyote or whether it is a narcotic or dangerous drug, it is difficult for them to determine the real or potential effects that the use of peyote, either by a few or by many, might have upon society.

The placing of peyote in the proper perspective with regard to U.S. society's value system is difficult, especially in the 1990s when cultural subgroups divided by age, ethnicity, and religion have widely divergent views on the acceptability of such substances as peyote. Society's acceptance of the practice seems to depend largely on the status of the drug user. Related to the controversy regarding the acceptability of a drug is the question of why humans have an urge to take drugs at all. In an article titled "Man the Drug Taker," this "desire to take medicine" is said to be "perhaps the greatest feature which distinguishes man from animals" (Osler, quoted in Bates 1967, 9). Bates added that "we seem to have no clear basis for determining which substances are drugs in the bad sense, except that they are substances socially disapproved of in our present culture" (9). Some drugs simply may never be accepted in a particular society because of certain undesirable physical effects, making laws pertaining to their use unnecessary. For example, millions chew the betel nut (*Areca catecheu*) in Asia, but its use is unlikely in the United States primarily because it stains the teeth and makes the person "unattractive."

Psychologists and philosophers have debated the subject of drugs extensively, and one interesting suggestion is that human beings have two powerful needs that seem to be at odds with each other: the need to keep things the same, and the need for change. To keep things predictable seems to be a function of one's ego; one wants to be able to control what happens in his or her life. But on the other hand, one often wants to rebel against this ego function and to find or do something with little or no predictability. This may involve simply a change of scenery, a change of heart, or a change of mind. But how do people change their minds toward the unpredictable? Many in our society turn to drugs—some to alcohol, but others to more "exciting" substances such as cocaine, marijuana, LSD, and heroin.

Of course, Americans have other "more acceptable" stimulants such as coffee and tea. Alcohol and tobacco have long been accepted, or at least tolerated, here, even though several scientists have stated flatly that tobacco is more unhealthy and habit-forming than marijuana, which is illegal. Such conflicts in society's laws can only bring about disagreements among subgroups and demands for change. The pressures of one group exerted on another within the same society—whether it be adults versus youths, Americans versus Native Americans, or Christians versus non-Christians—are especially evident when dealing with substances such as peyote and marijuana. Such practices are acceptable only so long as the overall welfare of society does not suffer—but no one seems to agree on what constitutes harm.

Some drug-taking experiences are described as highly beneficial, but so long as the results of drug-taking are unpredictable (and apparently that is why many people take them), there is potential harm not only to the individual but to society as well. This is the justification given for some form of social control.

In a 1970 article, physicians Henry L. Lennard, Leon J. Epstein, Arnold Bernstein, and Donald C. Ransom expressed deep concern regarding the hazards of using psychoactive drugs in determining "the context and the substance of our existence." They emphasized that "drug giving and drug taking represent all too brittle and undiscriminating responses, and ultimately, in our view, they will breed only more frustration and more alienation. Changing the human environment is a monumental undertaking. While seeking to change cognitive shapes through chemical means is more convenient and economical, the drug solution has already become another technological Trojan horse" (441).

The acceptability of a certain drug can only be determined on the basis of available information, but sometimes data are inadequate or invalid, and a law may be enacted that soon becomes either inappropriate or difficult to administer, especially when it runs counter to a centuries-old tradition.

Narcotics, Drugs, and Addiction

Views and opinions on some drugs are changing because people are learning more about them. In 1900 little was known about narcotics and addiction, and it is estimated that more than 4 percent of the U.S. population was addicted to narcotics that were common medicines and tonics of the day (see chapter 5). Physicians were largely unaware of the addicting properties of such substances as "Mrs. Winslow's Soothing Syrup" or "Perkins Diarrhea Mixture," but these harmless-sounding patent medicines usually contained opium or some equally potent narcotic, all freely available without a physician's prescription until the passage of the Harrison Narcotics Act of 1914. Enacted primarily as a method of narcotic control, this law made it necessary to obtain narcotics through physicians, and those who had become addicted were cut off from their easy supply of inexpensive drugs. Soon addicts were forced either to go through withdrawal or to obtain drugs illegally, and the act clearly led to a decrease in the number of users.

During the 1920s and 1930s the American value system changed as people learned about the existence of addiction, drug tolerance, and total dependence. Many people came to recognize that narcotics were no longer just simple "medicines" but dangerous substances, the use of which should be controlled by the government. But the public, and often government officials as well, lacked accurate scientific facts regarding various drugs and their addicting qualities and consequently drew false conclusions. Peyote, for example, was outlawed in California after a man was arrested on a street in Los

Angeles for selling heroin and peyote. The police and district attorney reasoned that since heroin was known to be a dangerous narcotic, peyote must be dangerous, too. Because of the lack of accurate information, the California legislature was persuaded to classify peyote as a narcotic, the definition of which is a major legal problem itself.

What is a narcotic? In the legal sense, a "narcotic" is any drug substance that can be classified as such by a legislature or law-making body. A scientific definition of "narcotic" would be most desirable, but until recently, no simple, or even complex, definition had been accepted among scientists and lawmakers. In its broadest classical definition, "narcotic" is based on the Greek word *narkōtikos* (which in turn is a derivative of *narkē*, meaning "numbness"), meaning "benumbing" or "producing sleep" (*Webster's Third New International Dictionary* 1981, 1503). Physicians and pharmacists now generally agree that a "narcotic" must be defined on the basis of its physiological action rather than its creation of a suspected (or actual) social problem (Schultes and Hofmann 1979, 11–12). There is also now general agreement that narcotics are the opioids and opioid antagonists (Goodman and Gilman 1970, 277), but any discussion of these substances seems to revolve around the physiological aspects of addiction and tolerance.

Addiction is the condition in which the individual has an overpowering desire or need to continue taking the drug, usually accompanied by a psychic (or psychological) and sometimes physical dependence on the effects that the drug produces (Expert Committee on Drugs Liable to Produce Addiction 1950, 6–7). In other words, the individual has the need to continue ingesting the drug because of the physical or psychological dependence it causes. Tolerance, discussed earlier (see chapter 6), often is closely associated with addiction. Tolerance develops when "after repeated administration, a given dose of a drug produces a decreasing effect or, conversely, when increasingly larger doses must be administered to obtain the effects observed with the original dose" (Goodman and Gilman 1970, 237).

Some people still contend that peyote is a narcotic despite statements to the contrary by experts with considerable knowledge of its physiological effects. Even early studies showed that peyote is not addicting. Maurice H. Seevers (1958, 244) calculated the "addicting liability" of a number of drugs found in U.S. society by constructing an index based upon the factors of tolerance, physical dependence, and habituation. He concluded, surprisingly, that alcohol was the most addicting, with an index factor of 21, followed by the barbiturates with an index of 18, opium and its derivatives (16), cocaine (14), and marijuana (8). Peyote, according to Seevers, had a liability of only one, due solely to the fact that some of his experimental subjects showed a slightly increased tolerance during the testing period.

This, then, raises an important question: Why is peyote repeatedly used by humans, particularly Native Americans? There is no compulsion to repeat the use of peyote because of addiction, withdrawal syndrome, or marked tolerance. Rather, the eating of peyote usually is a difficult ordeal in which nausea and other unpleasant physical manifestations occur regularly. Repeated use is unlikely, therefore, unless one is a serious researcher or is devoutly involved in taking peyote as part of a religious ceremony. The fact that curious users frequently comment that the ingestion of peyote was an "interesting experience" but not worth going through again seems to further validate the claims of many Native Americans that peyote is "medicine" and an important part of their culture.

The problem for lawmakers and scientists is one of deciding how best to classify and control the narcotics and other drugs used by humans. Marston Bates put his finger on the problem when he observed that our real concern

> is not so much in drugs taken as medicine to cure physical disease as it is in drugs used as "shortcuts to happiness"—drugs that dull the hard edge of reality, that produce excitement and heightened perception, or that lead the taker into a dream world where all is beauty and peace. These are the substances, in other words, that modify our mental state in ways that we find pleasurable or desirable.... We need a word that will cover everything from opium to tea. The best I can do is to call them all the pleasure-giving drugs—those taken for kicks—although many of them, in their primitive context, were serious enough, being connected with religious or magical operations. (Bates 1967, 9–10)

RELIGIOUS FREEDOM

The First Amendment of the Bill of Rights of the U.S. Constitution states that "Congress shall make no law respecting an establishment of religion, or prohibiting the free exercise thereof," but how can such religious freedom be insured when some perceive that it may cause harm to society? Conflicting views have forced lawmakers to decide whether the use of controlled or "dangerous" drugs in religious ceremonies is protected by the First Amendment, and the answer is not a simple one.

For nearly four centuries, lawmakers and other authorities have been divided over the question of the use of peyote for religious purposes. In the 1960s, 1970s, and 1990s, legislative and judicial bodies in the various states and in the federal government enacted laws and made decisions that clearly permit the religious use of peyote in the United States. Unfortunately, however, these regulations did not settle the issue; in 1990 the United States Supreme Court ruled that states may prohibit the use of peyote for religious purposes under certain circumstances (see discussion below). In response to this ruling, Congress in 1994 passed legislation that forbids the United States or any state from prohibiting the use of peyote by Indians for religious purposes.

The battles and disputes that preceded these rulings and legislative actions are an interesting facet of the story of the long use of peyote by Native Americans. The following accounts were selected because of their time in history and their ultimate effect on the legal situation regarding peyote.

The Spanish Conquest of Mexico in the sixteenth century apparently led to the first direct conflict of cultures concerning whether peyote could or could not be used ceremonially and religiously. This dispute quickly became a legal one because the Spanish Inquisition, in overseeing the moral wellbeing of all Christians according to Roman Catholic dogmas, possessed the power to enforce their proclamations through legislative, judicial, and police actions. It was apparently felt that the use of peyote presented a serious challenge to the Catholic Church, and in 1620 the Inquisition declared that the use of peyote was the "work of the devil" and that its use was prohibited. Irving Leonard, in a fascinating article titled "Peyote and the Mexican Inquisition," translated and quoted an edict of 1620 by Licenciado D. Pedro Nabarre de Ysla, in which the Inquisition ordered that

> henceforth no person of whatever rank or social condition can or may make use of the said herb, Peyote, nor of any other kind under any name or appearance for the same or similar purposes, nor shall he make the Indians or any other person take them, with the further warning that dis-

obedience to these decrees shall cause us, in addition to the penalties and condemnation above stated, to take action against such disobedient and recalcitrant persons as we would against those suspected of heresy to our Holy Catholic Faith. (Leonard 1942, 326)

When the peyote religion spread across the United States in the nineteenth century, Anglo religious leaders quickly became concerned and wanted to suppress its practice (see chapter 2). Because no state or federal laws existed to assist them in restricting the use of peyote, some missionaries to the Native Americans, other religious leaders, and various "well-doers" throughout the country attacked the religious use of peyote on two fronts: (1) by stretching the application of existing laws to include peyote, and (2) by appealing for state and federal laws to prohibit its use.

Examples of the former were the several attempts to include peyote as an intoxicant, thus putting it under the sway of various temperance pledges, prohibition laws, and even the Oklahoma Enabling Act (Marriott and Rachlin 1971, 44). Most such attempts proved legally futile, as several courts of law declared that the laws were specifically designed to deal only with alcohol intoxication (Collier 1937, 18241). The "temperance pledge" effort also proved only partially effective, because while there might be arrests for peyote usage, usually no sentences were pronounced because the law or pledge did not specifically mention peyote as an intoxicant (Stewart 1987, 128–47). Of course, many Native Americans strongly maintained that peyote and alcohol were completely different and that it was absurd to consider them to be the same.

The most effective actions were achieved by the passage of state laws prohibiting peyote, often involving widespread but often untrue newspaper publicity and other misinformation. In 1917, for example, a strong campaign was mounted against peyote in Colorado in order to obtain the necessary prohibitive legislation. The *Denver Post* of 12 January 1917 reported that the "societies which have interested themselves in the welfare of the Indians have discovered that peyote is killing dozens of Indians yearly. The 'peyote' eater has dreams and visions as pleasing as those of a 'hophead.' To get a better hold on their victims, the peyote peddlers have lent a religious tone to the ceremony of eating the drug, so that the peyote is worshiped in a semi-barbaric festival before the orgy is held" (Stewart 1956, 81). Denver organizations that supported a prohibitive measure against peyote included the Ministerial Alliance of Denver, the Women's Christian Temperance Union (WCTU), the Parent-Teacher Association, the Women's Club, and the Association of Collegiate Alumnae.

Such a response by well-meaning members of Anglo society is not surprising, because the peyote religion as reported to them seemed to be in direct opposition to established Christian mores. Further complicating the issue was the fact that the U.S. form of the peyote religion (as described in chapters 2 and 3) had taken on many Christian concepts and called itself the "Native American *Church*," thus creating a serious moral and legal dilemma: Should a "Christian church" that used "drugs" be permitted to function in our society? The First Amendment of the U.S. Constitution guarantees the free exercise of religion, but interpreting and applying this guarantee has long been a difficult legal problem. What precisely can be considered a "religion" has been continually open to question.

It is also important to note that none of the anti-peyote laws were passed as religious laws; they were enacted because legislators were convinced that taking peyote is harmful. These regulations have been the same as any other drug law: Use of the substance is harmful to the individual and to society; therefore, its sale, distribution, and use are controlled by specific legislation. Members of the Native American Church, who use peyote as their sacrament, argue that the plant is not dangerous and that the prohibition of its use is a violation of the First Amendment. But this is not a satisfactory answer to anti-peyotists who say that "it should not be legalized, even if only for religious purposes. If it's bad at home, it's bad at church" ("Button Eaters" 1959, 71).

Some state courts have sided firmly with those opposed to the religious use of peyote. In 1926 a Montana judge upheld the state's 1923 anti-peyote law and ruled that the use of peyote by Native Americans for religious purposes was not a valid reason to justify permitting anyone to eat it. His decision was based, in part, on the fact that although the state's constitution guarantees the free exercise and enjoyment of religion, it "also provides that the liberty of conscience thereby secured shall not be construed to dispense with the oaths or affirmations, excuse acts of licentiousness or polygamous marriage, or otherwise, or justify practices inconsistent with the good order, peace, or safety of the state" (Arthur 1947, 152). The judge concluded that the use of peyote was inconsistent with good order, peace, and safety of the state, *even if used for religious purposes.*

Thus, a serious legal issue was specifically raised with regard to the religious use of peyote: Are the effects of this plant harmful enough to deny a group of people their right of religious freedom? Opinions have been expressed on this question for nearly a century by anti-peyotists, who have been largely missionaries and other church workers with Native Americans. Missionaries have felt that peyote was more than just a problem involving the use of drugs with the related physiological and social ramifications. The use of peyote was religious, and worse yet, a revival of old traditional Native American customs and beliefs; it was a form of pagan worship that was very attractive to Native Americans for historical reasons and was highly competitive with Christianity. Newberne (1925), an outspoken opponent of peyote early in this century, stated emphatically that "to the missionary the use of peyote is paganism arrayed against Christianity—the power of a drug against the elevating influence of the cross" (8). Of course, he assumed that Christianity should be the religion of all people because of its superiority over any form of pagan religion, and that any Native American activity which ran against Christianity should be prohibited. He contended that peyotism, in mixing certain pagan beliefs with selected Christian elements, was an "absurd cult incompatible with Christianity" and opposed to the work of missionaries among the Native Americans (16). Missionaries were also concerned that peyotism involved a "vicious drug habit," an aspect that made it even more dangerous.

Anti-peyotists have also argued that it is the responsibility of the government to protect Native Americans and to integrate them into dominant European-immigrant ways. They have claimed that since it is an American principle to uplift downtrodden people such as Native Americans, aboriginal customs must be forbidden and Anglo culture imposed so that the indigenous people can rid themselves of ignorance and superstition (Bonnin and Bonnin 1937, 18297). Many opponents of peyote have believed that the protection of Native Americans from peyote is an essential part of governmental responsibility; the Bonnins and others argued that peyotism is one of the most serious threats to Native Americans because it "shackles" them to their pagan customs. The words of Congressman Carl Hayden reflected this view:

> The protection of the Indians against the intoxication produced by the peyote or mescal button is a matter of supreme importance. Not only does it affect the obligations which the Government has toward its wards,

but also particularly does it affect the physical, mental, and moral welfare of the Indians. . . . Intoxication produced by these bodies is of the most seductive character. . . .

The idea of making an intoxicating drug the basis of a religion is preposterous. One might as well use the sacrament as an excuse for drinking a gallon of wine to become intoxicated. This talk of religion is all a subterfuge. It is a bold attempt to perpetuate, under the guise of religion, the use of a drug that ought to be prohibited. (Hayden 1937, 18280)

Unfortunately, many anti-peyotists, including Congressman Hayden, have been badly misinformed and simply incorrect in some of their statements. Much of their "evidence" has been based on hearsay, and to a great extent they have been unwilling to listen to researchers in the various scientific fields. Even though most scientific evidence points to the fact that peyote does not affect sexual desires, anti-peyotists have repeatedly emphasized how the use of peyote, and the peyote ceremony itself, are sexually stimulating. Missionaries and other anti-peyotists have claimed that the peyotists participate in all-night "debaucheries" where often there is a "total abandonment of virtue, especially among the women" (Bonnin and Bonnin 1937, 18296). Accounts of crazed women tearing off their clothes and dancing almost naked have been circulated widely as "proof" that peyote is sexually stimulating. Even Native American opponents of peyotism have spread such stories, claiming that it "makes the men very bad" and that even the women "lose all of their ashamed" (Rave 1937, 18266). My own observations and those of other scientists and anthropologists indicate that peyote is not a sexual stimulant (see chapter 4).

In summary, opponents of the religious use of peyote have based their attacks on what they allege are its harmful effects on the body, its destruction of the moral fiber of Native Americans, and its attempt to masquerade as a genuine religion.

Supporters of the religious use of peyote, on the other hand, claim that few, if any, ill effects on users have been verified scientifically. They argue that peyotism *is* a valid religion and that the consumption of peyote is a significant element of the religious ceremony. They also contend that peyote is one of their most important "medicines," that it is not habit-forming, nor is it intoxicating. In fact, Native American peyotists strongly oppose the use of intoxicating substances, especially whiskey, which they believe leads to dishonesty and immoral habits. During congressional committee hearings about peyote in 1936 and 1937, Senator Elmer Thomas (1937) of Oklahoma, who had personally participated in peyote ceremonies, declared that the members of the peyote religion were the "more substantial Indians" of the Osage reservation in that they didn't drink, gamble, or use narcotics (18231). He added that most made some effort to work and did not appear to suffer adversely from continued use of peyote. Senator Thomas also commented that he had "never been convinced that peyote is any worse for the Indian than wine is for the whites, and I have never heard of an Indian using peyote except in connection with religious rites" (18167).

The same congressional committee read into its hearings an extensive study on peyote by Donald Collier of the University of Chicago (1937). Professor Collier's concluding statement summarizes the legal problem facing lawmakers.

> Peyote involves a practical question in the United States, which I have indicated but without suggesting an answer. It is the question of the governmental attitude toward peyote and its cult, and the prohibitive legislation, both State and Federal, which is a continual menace to the Indians' religion. I think I have made clear that there is nothing connected with the Indians which has been more misunderstood and less understood than the peyote cult. Therefore, I would urge an attitude of tolerance. Until more is known about the peyote cult, we have no right to suppress it. (Collier 1937, 18247)

More-recent research has supported Collier's plea for tolerance. Most investigations have shown that peyote is not a dangerous narcotic, as succinctly stated by psychiatrist Humphrey Osmond: "[A]ll the evidence that we have suggests that peyote is wholly beneficial and in no way a drug of addiction. It cannot even be defined in that way since it does not have the essential compelling qualities or the withdrawal symptoms" (quoted in Stewart 1956, 90). This is further supported by the work of Robert L. Bergman, a public health physician among the Navajo Indians. Dr. Bergman (1971) interviewed approximately two hundred Navajo peyotists and reported that he had "seen almost no acute or chronic emotional disturbance arising from peyote use" (697). Bergman further commented that the religious use of peyote seemed to be directed in an ego-strengthening direction with an emphasis on interpersonal relationships where each individual is assured of his own significance and the support of the group. He concluded that "we have seen many patients come through difficult crises with the help of this religion.... It provides real help in seeing themselves not as people whose place and way in the world is gone, but as people whose way can be strong enough to change and meet new challenges" (698).

In recent years an overall change in attitude by various governmental bodies toward the peyote religion has led to legislation specifically permitting the bona fide religious use of peyote by Native Americans. A governmental battle in New Mexico clearly shows that such laws have not come easily. In 1957, against the wishes of the Navajo Tribal Council, the New Mexico legislature passed a bill legalizing the religious use of peyote. It apparently was too hot a political issue and Governor Mechem promptly vetoed it. In 1959 the bill came before the state legislature again, and it was passed by a large majority (53 to 11 in the House) despite continued opposition from the Tribal Council ("Button Eaters" 1959). A new chief executive, Governor Burroughs, also opposed it, but he saw that a veto would do little good since there was sufficient legislative support to override his veto. He therefore allowed it to become law without his signature (Aberle 1991, 121).

Efforts to restrict the religious use of peyote were less successful at the federal level until the 1960s, when the increase in drug use began to cause many serious medical and social problems. Bills were introduced into Congress in 1916, 1917, 1918, 1921, 1922, 1924, 1926, and 1937, but none passed. The attempt of 1937, during the administration of Franklin D. Roosevelt, showed a continuing insensitivity towards Native Americans and their cultural differences during the extensive congressional hearings that were held. John Collier, director of the Office of Indian Affairs, who wanted to protect the religious freedom of Native Americans, suppressed a report by the Indian Bureau that strongly condemned the use of peyote; he then sent instructions to all tribes telling Native Americans that they could worship the way they wished. Collier's order produced such unfavorable reactions in Washington that in 1940 he was forced to back down and recommend approval of a strong anti-peyote law enacted by the Navajo Tribal Council. However, he emphasized that he recommended it only because he respected the authority of the council to govern the people of the reservation—but he would not specifically condemn the use of peyote as harmful (Stewart 1987, 295–97).

While some opponents of the peyote religion strove for federal control of the plant, others worked at the state level. Prior to the 1960s the latter groups were more successful and were instrumental in the passage of laws in eleven different states specifically prohibiting the use of peyote. Several of these anti-peyote laws have been challenged in the courts, both on the state and national level, as unconstitutional. Four court decisions bearing on this matter—in Arizona, California, Texas, and Oregon (which led to the significant United States Supreme Court decision of 1990)—merit further discussion.

The Arizona case occurred in 1960 when Mary Attakai, a Navajo, was arrested for illegal possession of peyote. She subsequently pleaded guilty in an Arizona court, but her attorneys argued that the law under which she was arrested violated the Fourteenth Amendment of the U.S. Constitution, as well as several sections of the Arizona Constitution with respect to insuring religious freedom. On 26 July 1960, Judge Yale McFate rendered his decision, commenting that few people used peyote except Native Americans practicing their religion and that there seemed to be nothing "debasing or morally reprehensible about the peyote ritual," of which peyote is an essential part. He declared that for all practical purposes Arizona's law prohibiting the use of peyote prevented worship by members of the peyote religion. In his formal statement, Judge McFate wrote:

> The manner in which peyote is used by the Indian worshipper is not inconsistent with the public health, morals, or welfare. Its use, in the manner disclosed by the evidence in this case, is in fact entirely consistent with the good morals, health and spiritual elevation of some 225,000 Indians.
>
> It is significant that many states which formerly outlawed the use of peyote have abolished or amended their laws to permit its use for religious purposes. It is also significant that the Federal Government has in no wise prevented the use of peyote by Indians or others.
>
> Under these circumstances, the court finds that the statute is unconstitutional as applied to the acts of this defendant in the conduct and practice of her religious beliefs. (*State of Arizona v. Mary Attakai*, Criminal Cause No. 4098, Coconino County, 1960)

The California case arose late in 1962 when law officers raided a peyote ceremony near Needles in which thirty persons were participating. Three Navajos were arrested and charged with violation of the state's narcotic law (peyote was classified as a narcotic in California at that time). A highly publicized trial followed in which numerous experts appeared as witnesses on behalf of the Native Americans. As in Arizona, the defense was based on the premise that the anti-peyote portion of the state narcotics law was unconstitutional because it infringed upon religious freedom. Judge Hilliard ruled that the defendants were guilty and asserted that religious freedom could not be used as an argument for the use of a drug classed as a narcotic in California. The district court of appeals upheld the lower court's decision, but the case was promptly passed to the California Supreme Court. After lengthy consideration, the state's highest court overturned the verdicts of the two lower courts (Stewart 1987, 309–10).

The significant decision, written by Justice Tobriner, noted that "the right of free religious expression embodies a precious heritage of our history. In a mass society, which presses at every point toward conformity, the protection of a self-expression, however unique, of the individual and the group becomes ever more important. . . . Law officers and courts should have no trouble distinguishing between church members who use peyote in good faith and those who take it just for the sensations it produces" (*People v. Woody*, 394 P.2d 813 [1963]). The justices emphasized that the use of the drug in "honest religious rites" was protected by the First Amendment of the U.S. Constitution. In a companion case the justices ruled that a person claiming religious exemption for possession of drugs must *prove* that his or her belief, which involves the use of peyote, is an "honest and bona fide one." Interestingly, no mention was made of whether the religious user of peyote needed to be a Native American.

In 1967 the Texas legislature passed a law forbidding the possession of peyote. With Texas being the only state where peyote occurs naturally, members of the Native American Church found that their legal supply of the plant was cut off. Frank Takes Gun, president of the Native American Church of North America, set up a test case in Webb County in which a young Navajo, David S. Clark, drove Frank Takes Gun's vehicle off the property of Amada Cardenas with peyote. He was arrested and the case went to court in 1968. Judge Kazen, presiding in the case, cited both the Arizona and California cases, as well as one in Colorado.

> The evidence in this case has shown that Peyotism is a recognized bona fide religion practiced by the members of the Native American Church, and that peyote is an essential ingredient of the religious ceremony; it is the sole means by which the members of the Church are able to experience their religion, and without peyote, the Court finds from the evidence, the members of the religion cannot practice their faith.

In view of the above evidence and findings, the Court finds and concludes that Article 726-d of the Penal Code of the State of Texas is unconstitutional as it applies to this defendant herein, who possessed and used peyote in good faith in the sincere and honest practice of Peyotism, a bona fide religion; and therefore, the defendant herein is found not guilty. (*State of Texas v. David S. Clark*, 1968)

In 1969 the Texas legislature amended the penal code to provide the following exemption:

The provisions of this chapter relating to the possession and distribution of peyote do not apply to the use of peyote by a member of the Native American Church in bona fide religious ceremonies of the church. However, a person who supplies the substance to the church must register and maintain appropriate records of receipts and disbursements in accordance with rules adopted by the director. An exemption granted to a member of the Native American Church under this section does not apply to a member with less than 25 percent Indian blood. (HSC Sec. 481.111)

It is important to note that the Texas legislature stated that a person was considered to be an Indian only if he or she had at least 25 percent Indian blood.

Many people have feared that such rulings as those in Arizona and California would open the door to new churches that use drugs. Indeed, several organizations using drugs in a manner similar to the peyote church have arisen. An example is the Church of the Awakening, founded in 1963 in New Mexico by two retired osteopaths, John and Luisa Aiken. The church's charter emphasized "the search within one's own consciousness for the Self, which is Being, which is Life" (Braden 1967, 118). Hallucinogenic drugs were served as the sacrament. This organization had little success in receiving exemption for religious purposes from the federal dangerous-drug laws (Stewart 1987, 326). Other churches professing similar beliefs are the Neo-American Church and the League for Spiritual Discovery.

THE COMPREHENSIVE DRUG ABUSE PREVENTION AND CONTROL ACT OF 1970

Virtually all efforts to prohibit peyote on the federal level were unsuccessful until 1965, when amendments to the Food, Drug, and Cosmetic Act modified the definition of "Depressant and Stimulant Drugs" to include "any drug which contains any quantity of a substance which the Secretary, after investigation, has found to have, and by regulation designates as having, a potential for abuse because of its depressant or stimulant effect on the central nervous system or its hallucinogenic effect" (79 STAT 226). By adding the hallucinogens, which included peyote, peyote was therefore prohibited. Five years later, peyote and the other hallucinogens were dealt with in a similar manner in the Comprehensive Drug Abuse Prevention and Control Act of 1970. Building upon the 1965 amendments to the Food, Drug, and Cosmetic Act, this far-reaching legislation has led most states and the federal government to legally classify peyote as a hallucinogen.

According to pharmacologists Carl C. Pfeiffer and Henry B. Murphree (1965), hallucinogens are substances that "mimic the major psychoses as they occur in man" (324). Stephen Szara (1972) divides the hallucinogens into three chemical groups: (1) the phenylethylamine group, which includes mescaline and other epinephrine-related substances, (2) the tryptamine group, which includes LSD, psilocybin, and related compounds, and (3) a heterogeneous group of non-nitrogenous compounds that have somewhat different psychic and vegetative effects on man, such as loss of contact with the environment and amnesia during the period of hallucinations. Two examples of this last group would be myristicin, the active substance of nutmeg (*Myristica fragrans*), and the tetrahydrocannabinol alkaloids of *Cannabis sativa* (marijuana).

Some of the other names that have been applied to the hallucinogens are psychotomimetic (= mimicking a psychosis), psycholytic, psychedelic (= mind-manifesting), phanerothyme, phantasticant, schizophrenogenic, psychodysleptic, deliriant, mysticomimetic, misperceptinogen, delusionegen, psychoticant, psychogen, psychotogen, dysleptic, entheogen, psychotaraxic, eidetic, schizogen, and cataleptogenic. Many arguments have been made expressing a preference for one term over another, usually because the person wishes to emphasize only one of several manifestations of the drug's action. The term "hallucinogen," for example, stresses perceptual changes including the production of hallucinations, whereas "psychotomimetic" emphasizes the mimicking of a psychosis or a near-psychotic condition.

Richard Evans Schultes and Albert Hofmann, in an attempt to include the major features of this group of drugs in a meaningful definition, have proposed a description that is appropriate for the substances under consideration. They state that hallucinogens are

> agents which, in nontoxic doses, produce, together or alone, changes in perception, thought and mood, without causing major disturbances of the autonomic nervous system. A variety of hallucinations may be characteristic, especially with high doses. Disorientation, loss or disturbance of memory, excessive impairment of intellectual powers, hyperexcitation or stupor or even narcosis may be experienced only under excessive doses and cannot, therefore, be considered characteristic. Addiction is unknown with these drugs. (Schultes and Hofmann 1980, 15)

Although the hallucinogens do not form a single chemical group, all stimulate the peripheral sympathetic nervous system and greatly affect the senses, particularly the visual.

Peyote use by non-Indians is still regulated under the Comprehensive Drug Abuse Prevention and Control Act of 1970 (Public Law 91-513), which has superseded all other federal legislation dealing with peyote as a drug and established five schedules of controlled substances, to be known as Schedules I, II, III, IV, and V. Hallucinogenic substances, which include peyote and mescaline, are classed as Schedule I substances. Also included in this schedule are opium, LSD, marijuana, and psilocybin. Schedule I substances are defined as having a "high potential for abuse," no currently accepted medical use in the United States, and a lack of "accepted safety" under medical supervision (84 STAT 1247).

Section 404 of the act states that no person can knowingly or intentionally "possess a controlled substance" unless obtained by a prescription or "while acting in the course of his professional practice." Therefore, simple possession of either peyote or its alkaloid mescaline is unlawful and violators are subject, if convicted, to "imprisonment of not more than one year, a fine of not more than $5,000, or both" (84 STAT 1264).

Of particular significance is section 21 CFR 1307.31 of the law, which specifically exempts members of the Native American Church.

> The listing of peyote as a controlled substance in schedule I does not apply to the non-drug use of peyote in bona fide religious ceremonies of the Native American Church, and members of the Native Ameri-

can Church so using peyote are exempt from registration. Any person who manufactures peyote for or distributes peyote to the Native American Church, however, is required to obtain registration annually and to comply with all other requirements of the law.

The Comprehensive Drug Abuse Prevention and Control Act clearly distinguishes the hallucinogens—mescaline, peyote, LSD, psilocybin, and marijuana—from the true narcotics, which are defined in the law as opium, coca leaves, and opiates. However, peyote and mescaline are still classed as Schedule I controlled substances, and possession is prohibited unless specific permission was obtained or one is a member of the Native American Church. Permits can be secured from the U.S. attorney general's office if any of the controlled substances are needed for teaching or research. The application, which requires a research protocol and a description of security procedures, is sent directly to Washington, D.C. Federal law prohibits the possession of peyote plants by cactus collectors and horticulturists. Persons who collect peyote for religious purposes are required to be licensed (Sec. 307.31).

The federal Comprehensive Drug Abuse Prevention and Control Act of 1970 has stimulated many state legislatures to bring their own laws into agreement with the federal one, and in most cases peyote and marijuana have come to be called hallucinogens, not narcotics. Twelve states (Alaska, Mississippi, Montana, New Jersey, North Carolina, North Dakota, Rhode Island, Tennessee, Utah, Virginia, Washington, and West Virginia) have had laws directly related to the federal law concerning the use of peyote for religious purposes. Three states (Colorado, Nevada, and New Mexico) give full exemption for any bona fide religious use. Seven states (Iowa, Kansas, Minnesota, South Dakota, Texas, Wisconsin, and Wyoming) have full exemption for bona fide religious use by the Native American Church. Two states (Arizona and Oregon) consider bona fide religious use of peyote as an affirmative defense. The State of Idaho exempts peyote for bona fide religious use only on reservations. California and Oklahoma have court-created exemptions for the religious use of peyote. However, as will be shown in a following section, federal legislation now prohibits any state from prosecuting Native Americans when they use peyote for religious purposes.

THE AMERICAN INDIAN RELIGIOUS FREEDOM ACT OF 1978

In 1978 the United States Congress passed the American Indian Religious Freedom Act, which became Public Law 95-341. This law says in part that "it shall be the policy of the United States to protect for American Indians their inherent right of freedom to believe, express, and exercise the traditional religions of the American Indian, Eskimo, Aleut, and Native Hawaiians, including but not limited to access to sites, use and possession of sacred objects, and the freedom to worship through ceremonials and traditional rites" (42 U.S.C. 1996).

For many years, members of Congress had recognized the conflict of cultures between Native Americans and European immigrants. This act was an attempt to prevent encroachments upon the religions of Native Americans and to insure that they could freely exercise their religious freedom. The act cites insensitivity and ignorance as two significant factors causing encroachments upon Native American religious practices (Loftin 1989, 30). The American Indian Religious Freedom Act sought to correct several problems involving Native American religious practices, such as collection of eagle feathers and the protection of sacred sites. Thus far, most of the suits brought under the act have dealt with sacred sites. Unfortunately, the law has not been effective, because enforcement procedures are absent. Also, the act has tended "to give too much weight to the establishment clause" and "to stress differences over relationships" (Loftin 1989, 31). The act has not been interpreted to include protection for the Native American Church and its sacramental use of peyote. This became very evident with the United States Supreme Court's action of 1990 (see appendix C for a more complete text of the law).

THE UNITED STATES SUPREME COURT DECISION OF 1990

One of the most far-reaching and significant court decisions involving peyote began in almost an insignificant way. It involved whether a person could be denied unemployment benefits for violating a state law prohibiting the use of peyote, despite the fact that it was used for religious purposes. It was not a criminal case, and no one was ever arrested. In Oregon it is a crime to use or possess peyote, and no formal provision is made in the law for members of the Native American Church to use the cactus ceremonially, as do the other western states (Tepker 1991, 2). Two members of the Native American Church worked at an Oregon drug rehabilitation center as counselors in 1984. One of the men, Al Smith (1993), related that

> They called me into the office on Friday because I told them that I would be going to a Tipi Meeting. They called me and asked me if I attended that meeting, that I was not to ingest that mind altering drug Peyote, and again I tried to explain to them, I says to them, it's not a drug. For native people of the Native American Church, it's a very, very sacred sacrament.
>
> They called me into the office on Monday, and asked me if I'd attended the Native American Church ceremony, and I said I did. And they asked me if I ingested that mind altering drug Peyote, and I says no, but I did take the sacrament, the sacred sacrament, I says and they says you leave us no alternative, but we have to terminate you, ya know. It's always been, always been a misunderstanding of cultures, misunderstanding of definitions. It's been a language, language misunderstanding.

The two Native Americans were terminated from their jobs, and they subsequently applied for unemployment compensation. Though they had not been charged with criminal behavior with regard to the peyote incident, both were denied unemployment compensation on the basis that they had been fired for *criminal* "misconduct."

The two cases were appealed to the Oregon Court of Appeals and then the Oregon Supreme Court. Both courts ruled in favor of the defendants, stating "that the denial of benefits violated respondents' rights under the first amendment to free exercise of religion" (Tepker 1991, 3). However, the full analysis of the constitutional issues was argued only in the Smith case. The attorney general of Oregon therefore appealed only the Smith case to the United States Supreme Court.

The United States Supreme Court twice considered the Smith case, known as *Employment Division, Department of Human Resources of Oregon v. Smith* (485 U.S. 660 [1988] and 110 S.Ct. 1595 [1990]). First, the Supreme Court returned it to the state, ruling that Oregon had to explain specifically whether or not the sacramental use of peyote was a violation of state law. Otherwise, to rule on whether the rights of the First Amendment had been violated was premature. The Oregon Supreme Court found that

Oregon law made no exception for the "sacramental use" of peyote; thus, the two Native Americans had, indeed, violated Oregon's criminal law dealing with controlled substances (Sec. 80-005 of Chapter 855, defining "dangerous drugs" as provided in ORS 475.100 and 475.110). They had not been arrested for violating this law, despite the fact that they admitted they had used peyote. However, the Oregon Supreme Court *again* ruled that Smith was entitled to unemployment benefits. The state attorney general appealed once again to the United States Supreme Court, and it agreed to hear the case.

The United States Supreme Court's 17 April 1990 decision reversed the Oregon court's ruling by a vote of 6-3: Smith was not entitled to unemployment compensation. Justice O'Connor voted with the majority to reverse the Oregon Court's decision, but her rationale differed from that of the majority. Biskupic (1991), writing a summary of the case, said that the Supreme Court's decision was that "a state does not have to prove that it has a 'compelling interest' in enforcing a statute that happens to infringe on religious practice. . . . As long as the law is reasonable and religious burden is not its main purpose, the First Amendment is not breached" (916). Justice Scalia wrote:

> If the "compelling interest" test is to be applied at all, then, it must be applied across the board, to all actions thought to be religiously commanded. Moreover, if "compelling interest" really means what it says (and watering it down here would subvert its rigor in the other fields where it is applied), many laws will not meet the test. Any society adopting such a system would be courting anarchy, but that danger increases in direct proportion to the society's diversity of religious beliefs, and its determination to coerce or suppress none of them. Precisely because "we are a cosmopolitan nation made up of people of almost every conceivable religious preference," *Braunfeld v. Brown,* 366 U.S. at 606, 81 S.Ct. at 1147, and precisely because we value and protect that religious divergence, we cannot afford the luxury of deeming *presumptively invalid,* as applied to the religious objector, every regulation of conduct that does not protect an interest of the highest order. (110 S.Ct. at 1605)

In other words, the majority of justices ruled that individual states may enforce their criminal laws even if such enforcement infringes on the religious liberties of a minority within that state. Justice Scalia says "that unavoidable consequence of democratic government must be preferred to a system in which each conscience is a law unto itself or in which judges weigh the social importance of all laws against the centrality of all religious beliefs" (110 S.Ct. at 1606).

The United States Supreme Court's decision in the Smith case involved two distinct rulings. First, by a vote of 6-3, it reversed the Oregon court's ruling and ruled that the state can enforce a law that indirectly infringes on religious freedom. Thus, a state *can* outlaw the religious use of peyote, *though it does not have to.*

Second, by a 5-4 majority, with Justice O'Connor shifting to join the minority, the Court ruled that the state does not have to prove "a compelling state interest" in enforcing a law that might indirectly infringe upon one's First Amendment guarantee. Justice O'Connor objected to this ruling, pointing out that it "departed from well-settled First Amendment jurisprudence." The "compelling state interest" had been the standard since 1963. Thus, in the Smith decision, the court established a new standard: As long as the law is reasonable and its prime target is not religion (it is not a burden to religion), then the First Amendment is not breached.

The effect of this ruling, seen in light of the country's so-called war on drugs, is that there can be compelling state and federal interest to prohibit the use of peyote, even by those who use it for religious purposes. Thus any state can regulate the use of peyote by the Native American Church "in an absolute fashion. . . . Here state law overpowers previous federal policies, congressional intent, and presumed federal constitutional guarantees of First Amendment protection" (Lawson and Morris 1991, 82–83).

The Supreme Court's decision released a firestorm of protest. Fifty-four constitutional scholars and fifteen organizations, many from the religious community, petitioned the court to reconsider its decision (Laycock 1990, 519), but it refused to do so.

Congressional Actions Since 1990

Within the past few years, several bills have been introduced into Congress to amend or replace the American Indian Religious Freedom Act.

Congress enacted the Religious Freedom Restoration Act in 1993 as a direct response to the United States Supreme Court ruling of 1990. In effect, this act reinstated the "compelling government interest" test in determining whether governments were infringing on religious freedom (see appendix C for pertinent text of this act). However, it still remained unclear whether the traditional use of peyote by the Native American Church was protected.

In 1994 an amendment to the American Indian Religious Freedom Act known as H.R. 4230 was proposed in both houses of Congress to specifically protect the religious use of peyote. It was passed by voice votes in both houses and signed into law by President Clinton, becoming Public Law 103-344 on 6 October 1994 (see appendix C). For the first time, there is now specific federal legislation permitting the religious use of peyote by Indians throughout the United States. Indians, defined by the amendment as "a member of an Indian tribe," can no longer be prosecuted in any state for transporting, possessing, or using peyote for bona fide traditional ceremonial purposes in connection with the practice of a traditional Indian religion.

Legislation concerning peyote and similar substances has varied greatly within the United States, but the federal Comprehensive Drug Abuse Prevention and Control Act of 1970 provided the basis for revisions and standardization of most state laws. All states consider peyote to be a controlled substance and prohibit its possession. On the other hand, bona fide religious use of peyote is now specifically permitted by federal law, despite the United States Supreme Court ruling of 1990 which said that "compelling state interest" could prohibit members of the Native American Church from practicing their religion. This may or may not settle the issue, one that has persisted for hundreds of years, fed in large part by cultural ignorance and misunderstanding.

Perhaps times are changing and greater respect will be shown to Native American cultures and religious practices. The use of peyote as a sacrament is one of the oldest. However, persecution and harassment may continue, despite federal legislation that now exists to protect members of the Native American Church and their sacramental use of peyote. If future legal controversy occurs, lawmakers and the judiciary would be well advised to study the large amount of scientific evidence and not be misled by the misinformation and emotionalism that seem to have accompanied many previous governmental studies of peyote.

The Dívine Cactus

Over the past several years I have made numerous trips to the desert regions of south Texas and northern Mexico. Frequently, I have paused to look at peyote plants half-hidden by agave leaves, debris, and sand, and remembered the first time I saw peyote nearly forty years ago. How insignificant it had looked to me at that time! I remember wondering why anyone even paid any attention to it. After all, it was a relatively insignificant little cactus spread over a wide area of desert. Over the years, however, I slowly began to realize that this plant, though biologically only a minor component of the widespread Rio Grande Plains and Chihuahuan Desert, has perhaps influenced Native American cultures as much as any other New World plant. The simple fact that Native Americans discovered peyote and its remarkable effects upon the mind emphasizes how familiar early humans were with the flora and fauna of the North America, for their existence depended on their knowledge of the natural world around them.

This small and insignificant-appearing cactus means more to some of the indigenous people of Mexico and other parts of North America than does maize, their everyday source of food, for peyote goes beyond the material things of daily human existence and into the spirit world. Through peyote many Native Americans are able to reach out of their physical lives, to communicate with the spirits, and to "become complete."

Peyote has interested many scientists and scholars. Historians and anthropologists have sought out the story of the peyote religion and how humans have integrated this plant into their cultures over the generations. Chemists, pharmacologists, and psychiatrists have investigated the alkaloids present in peyote and have studied the ways in which they work upon the human mind. Physicians have attempted to learn if peyote is, indeed, the important "medicine" that many Native Americans believe it to be.

What might the future hold for peyote? How will people deal with it in the years to come? Change is certainly occurring, some good, but some discouraging, particularly with regard to the long-term survival of some natural populations. Thirty years ago I found peyote growing plentifully on many hillsides in the desert area of the Mexican state of Querétaro, but within a few years many were removed. I was told that commercial cactus collectors had come to the area and had hauled away thousands of plants. Some of the peyote gardens of south Texas are under severe pressure from the legal collection of tops by Native American pilgrims and by peyoteros for sacramental use in the Native American Church. This crisis in Texas is exacerbated by the closing of ranches by the landowners to both Native American Church members and peyoteros, as well as by the destruction of natural vegetation, including peyote plants, by root plowing to facilitate cattle grazing. Thus, there exists a dilemma: clearly, protection is needed for peyote from illegal collecting, but at the same time it is important that peyote be available for members of the Native American Church. Part of the solution may be the salvaging and cultivation of peyote.

Laws concerning peyote reflect an increasing awareness of nonindigenous people concerning the cultural and religious needs of Native Americans. Though both the U.S. and Mexican governments prohibit the collecting or possession of peyote by those who are not Native Americans, both governments are demonstrating increased tolerance to the religious use of the plant by the indigenous people. Thus it seems the use of peyote in Native American religious practices should be safe from persecution in the future.

On the other hand, legislation in Mexico and the United States has placed restrictions on the possession of peyote by nonindigenous people, including scientists, whose research is permitted in the United States only by special federal permit. Because possession of peyote is illegal, this interesting cactus is absent from most private cactus collections and botanical gardens within the United States and Mexico. Interestingly, there are no laws against possession of peyote in Europe. In fact, I have even seen peyote plants for sale at an open market in front of the cathedral in Cologne, Germany!

Peyote holds a profound place in the cultures of many groups of Native Americans. As Albert Hensley, the great Winnebago peyote leader and missionary, so clearly stated: "Our favorite term [for peyote] is 'Medicine,' and to us it is a portion of the body of Christ, even as the communion bread is believed to be a portion of Christ's body by other Christian denominations. . . . It came from God. It is a part of God's body. God's Holy Spirit enveloped in it. It was given exclusively to Indians and God never intended that White men should understand it, hence the folly of any such attempt" (quoted in Stewart 1987, 157). Though Hensley felt that many of us could never understand peyote and its significance, I am nonetheless hopeful that, even though we may never understand it as do Native Americans, this book will help us gain a greater appreciation and respect for this remarkable cactus. Perhaps it will also help eliminate some of the prejudices and untruths that have persisted about peyote and the people who revere it.

Appendix A : Peyote Systematics

The following formal botanical treatment of peyote is based on my 1969 article "The Biogeography, Ecology, and Taxonomy of *Lophophora* (Cactaceae)," as well as additional research since that time. This appendix includes the taxonomy of peyote: a generic description of *Lophophora*, its distribution, a key to the two species of *Lophophora*, descriptions and distributions of the species, a listing of representative specimens from major North American herbaria, and an extensive discussion of the synonymy. As mentioned in chapter 8, peyote has at one time or another been included in several different genera, and numerous species and varieties have been described within the genus. It is therefore important to discuss and evaluate the nomenclature of peyote in order to clarify my present conclusion that there are two species: *Lophophora williamsii* and *L. diffusa*.

Citation of herbaria follows the system of *Index Herbariorum* (Holmgren, Keuken, and Schofield 1981).

Taxonomy of *Lophophora*

LOPHOPHORA J. Coulter, Contrib. U.S. Natl. Herb. 3: 131. 1894.
TYPE: *L. williamsii* (Lemaire) J. Coulter.
Plants low and with long fusiform roots. Stems solitary or in clusters arising from the same root system, usually rounded above, depressed in the centers; blue-green, yellow-green, and occasionally appearing reddish-green; 2–7 cm high, 4–12 cm in diam. Areoles usually linearly arranged along ribs or at the apices of hump-like tubercles or podaria, each bearing a tuft of soft, yellowish or whitish trichomes from which the flowers arise, 8–15 mm apart, 1–5 mm in diam; spines absent except in seedlings and then only rudimentary. Flower 1–2.2 cm in diam, 1–2.4 cm long; outer perianth segments with greenish midribs and greenish-pink or whitish margins, the largest ones elliptical, 3–12 mm long, 1–3 mm broad, mucronate, marginally minutely ciliate distally; inner perianth segments pink, pinkish-red, white, or rarely yellowish-white, sometimes with greenish midribs, the largest ones elliptical, 8–22 mm long, 2–4 mm broad, mucronate or occasionally attenuate, margins ciliate or entire; filaments white or rarely magenta; anthers yellow, pollen ± spheroidal, 0–18 colpate, 26–63 mm diam; style white or rarely magenta, 5–14 mm long; stigmas 4–8 (rarely 3), 1–3 mm long, white though occasionally pinkish; ovary naked. Fruit pinkish-red and fleshy at first, becoming brownish-white and dry at maturity, naked, clavate, 15–20 mm long, 2–3.5 mm in

diam, with the umbilicus large or the perianth parts persistent, emerging rapidly from within the trichomes at maturity. Seeds black, tuberculate (verrucose), pyriform, 1–1.5 mm long, 1 mm broad; hilum large and flattened; cotyledons coalescent.

Flowering from March through September. Fruits mature 9–12 months after fertilization.

Widely distributed on limestone soils of low hills and flatlands in the Chihuahuan Desert and Rio Grande Plains of Texas, and the Chihuahuan Desert of northern and central Mexico, from 50 to 1800 m elevation.

Lophophora is derived from two Greek words meaning "I bear crest," referring to the "crests" of trichomes borne on each tubercle. Common names are peyote, piote, piotl, peyotl, peyori, pezote, pejote, peyot, pellote, piule, peote, challote, mescal, mescal button, divine cactus, devil's root, diabolic root, raíz diabólica, dry whisky, dumpling cactus, cactus pudding, turnip cactus, white mule, Indian dope, moon, "P," the bad seed, and tuna de tierra. There are also numerous names in Native American languages.

KEY TO THE SPECIES

A. Plants blue-green, usually with well-defined ribs and furrows; tufts of trichomes usually equally spaced on the ribs; flowers pinkish or rarely whitish; not in Querétaro. . . . . 1. *L. williamsii*
AA. Plants yellow-green, usually lacking well-defined ribs and furrows; tufts of trichomes usually unequally spaced on prominent podaria; flowers commonly whitish to yellowish-white; Querétaro. 2. *L. diffusa*

1. *L. WILLIAMSII* (Lemaire) J. Coulter, Contrib. U.S. Natl. Herb. 3: 131. 1894.
NEOTYPE: On the flat lands northeast of the junction of federal highways 57 and 80 near El Huizache, San Luis Potosí, 3 July 1958, *E. F. Anderson 1079*, POM.

Plants somewhat firm, blue-green and occasionally reddish-green often ± flattened on top, 2–6 cm high, 4–12 cm in diam; ribs and furrows usually present and well defined, 4–14 (and occasionally more) but highly variable and sometimes forming ± elevated podaria; areoles 0.9–1.5 cm apart, 2–4 mm in diam; flower 1–2.2 cm in diam, 1–2.4 cm long; outer perianth segments 2–3 mm broad; inner perianth segments 2.5–4 mm broad, usually pink; pollen 0–18 colpate, 14.9–63.4 mm in diam (fig. A.1).

Occurring in Texas, south of Shafter, in Big Bend National Park, at the mouth of the Pecos River, and from Laredo south-eastward to McAllen. In Mexico, in the state of Coahuila southward from the border west and north of Saltillo, extending

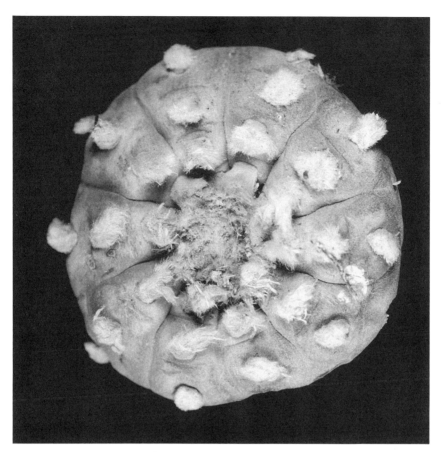

FIGURE A.1. *Lophophora williamsii* (Lemaire) J. Coulter.

eastward and southward into Nuevo León, Tamaulipas, northeastern Zacatecas, and San Luis Potosí (see fig. 8.2).

The specific epithet is derived from the proper noun, but it is not known to whom the plant was dedicated by Lemaire.

Representative specimens:
UNITED STATES: TEXAS: Presidio Co.: E of Shafter, *B.L. Warnock s.n.* (SRSC); S of Shafter, *N.H. Boke s.n.* (POM); S of Shafter, *E.F. Anderson 925, 2308* (POM). Brewster Co.: Chilocotal Mt., Big Bend Natl. Park, *B.L. Warnock 18498* (SRSC). Val Verde Co.: Mouth of the Pecos River, *W.M. Lloyd s.n.* (US). Webb Co.: Laredo, *Weinberg 25477* (NY); Laredo, *Shiner s.n.* (US); Laredo, *B. MacKensee s.n.* (US). Starr Co.: N of Rio Grande City, *E.U. Clover 1879* (DS, MEXU); N of Rio Grande City, *R.O. Albert s.n.* (POM); E of Rio Grande City, *J.N. Rose 24374*

(US); N of Rio Grande City, *E.F. Anderson 1127* (POM); E of Rio Grande City, *E.F. Anderson 1129* (POM); N of Escobares, *E.F. Anderson 947, 1126* (POM). Jim Hogg Co.: S of Mirando City, *E.F. Anderson 948, 1123, 2307* (POM).

MEXICO: TAMAULIPAS: S. of Reynosa, *E.F. Anderson 1134* (POM); W of La Perdida, *E.F. Anderson 2275* (POM); W. of Miquihuana, *E.F. Anderson 1726* (POM). COAHUILA: N of Saltillo, *E.F. Anderson 1063, 2306* (POM); Saltillo, *E. Palmer 688, 439* (US); Saltillo, *E. Palmer 94* (US, GH, MO); Near Saltillo, *E. Palmer 401* (US); W of Saltillo, *E.F. Anderson 1238 2305* (POM); E of Saltillo, *E.F. Anderson 1250, 2304* (POM); Cerro del Zapatero, *C.A. Purpus s.n.* (US); SW of Parras, *E.F. Anderson 1248* (POM). NUEVO LEON: N of Matahuala, *E.F. Anderson 1228, 2303* (POM). SAN LUIS POTOSI: al sur de Loma Bonita, *H. Sánchez-Mejorada 2607* (MEXU); O de Nuñez; *J. Rzedowski 5796* (SLPM); 30 km S del campo "La Sauceda," *L. Scheinvar 2320* (MEXU); KM 61 de Matehuala a San Roberto, *H. Sánchez-Mejorada 4027* (MEXU); SO de El Cedral, *H. Sánchez-Mejorada 2104* (MEXU); W of El Huizache jct., *E.F. Anderson 1211* (POM); Entronque carreteras 57 y 80, *A. Gomez G. 55* (SLPM); El Huizache jct., *E.F. Anderson 1079* (POM, US, NY, MO, GH, K); El Huizache jct., *E.F. Anderson 961, 1256, 1757, 2287* (POM); E of El Huizache, *R.D. Bratz s.n.* (POM); E of El Huizache, *E.F. Anderson 1215* (POM); 1 km adelante del Huizache, *L. Scheinvar 4905, 4907* (MEXU); Near Las Tablas, *E.F. Anderson 1752* (POM); W of Ciudad del Maíz, *E.F. Anderson 1182, 1615, 2280* (POM). ZACATECAS: Northern Zacatecas, *F.E. Lloyd 52* (US, NY); K200 de la carretera Zacatecas a Saltillo, *H. Bravo H. 49* (MEXU).

Echinocactus williamsii Lemaire ex Salm-Dyck in Otto and Dietr., Allgem. Gartenzeit. 13: 385. 1845, *Anhalonium williamsii* Ruempler in Förster, Handb. Cact. ed. 2, 233. 1886, *Mammillaria williamsii* J. Coulter, Contrib. U.S. Nat. Herb. 2: 129. 1891, *Ariocarpus williamsii* Voss, Vilmorin Illustrirte Blumengärtneri 368. 1894, and *Lophophora williamsii* J. Coulter, Contrib. U.S. Nat. Herb. 3: 131. 1894. The original epithet was attributed to Lemaire by Salm-Dyck, who stated that the original name appeared in Cels Catalog in 1845 without a description. Salm-Dyck presented a Latin description, but did not mention the origin of the plant or its type locality. No illustration accompanied Salm-Dyck's description; apparently the first illustration of this plant appeared as a color plate in Curtis's Bot. Mag. (pl. 4296) two years later. As no plant was designated as the type specimen and apparently there are no specimens preserved in Europe by Lemaire or Salm-Dyck, the following specimen is designated as a neotype: "On the flat lands northeast of the junction of federal highways 57 and 80 near El Huizache, San Luis Potosí," *Edward F. Anderson 1079* 3 July 1958. Neotype, POM 298103.

Echinocactus rapa Fischer and C. Meyer ex Regel, Sertum Petropolitanum,

Plate 13. 1869, with description. This binomial was unknown to most botanists because of the rare publication in which it occurred until Fred R. Ganders (1975, 155–156.) rediscovered it and published an article in which he reproduced the illustration and description of *E. rapa*. He was correct in identifying the pictured cactus as *L. williamsii*, which is the third oldest illustration of peyote. Because it is a later homonym, *E. rapa* becomes a synonym of *L. williamsii*. No type specimen was mentioned.

Anhalonium lewinii Hennings, Gartenflora 37: 410. 1888, *Echinocactus lewinii* Schumann in Engler & Prantl, Nat. Pflanzenfam. 3, 6a: 173. 1894, *Lophophora williamsii* var. *lewinii* J. Coulter, Contrib. U.S. Nat. Herb. 3: 131. 1894, *Lophophora lewinii* Rusby, Bull. Pharm. 8: 306. 1894, and *Mammillaria lewinii* Karsten, Fl. Deutsch. ed. 2, 2: 457. 1895. This epithet was proposed by Hennings based on body form and the characteristics of the tufts of hair at each areole. He stated that the hairs of the Salm-Dyck species were whiter, silkier, and longer than those of *A. lewinii*. The latter species was also said to possess a larger hairy pad in the center of the plant and had some differences in flower parts. Henning's description and illustration were based on dried plant material received under the designation "Muscale buttons" by Louis Lewin in 1887 from Parke, Davis, and Company in the United States. From the same materials Lewin extracted alkaloids and studied their physiological effects. It was Lewin's belief that only *A. lewinii* possessed the alkaloids, but Todd's study of 1969 shows all peyote plants to contain them (see chapter 7). Another illustration of *A. lewinii* was later published by Arendt in 1891; this showed that the body was not as flat and had more trichomes than did the Salm-Dyck species. Body shape and quantity of trichomes are highly variable in *Lophophora* and it would be impractical to base a species on these characters alone, especially when the original description and illustration were based on dried rather than living cactus material. Specimens of peyote are soft and succulent, and there is little cork or woody tissue of any kind. A dried specimen has a different shape, and the relative amount of surface covered by trichomes is altered because the trichomes maintain the same volume but the body decreases in volume through loss of water. Consequently, a dried specimen has a more woolly apical area than the same specimen had when living. Somewhat later Ochoterena (1922, 97) distinguished *A. lewinii* on the basis of its flower color and lack of definite ribs; he reported the flowers of *A. lewinii* to be yellow, but Henning's original description clearly states that the flowers were pink. Absence of or number of ribs has also been a commonly used diagnostic feature, but field studies show that single clones have both ribbed and non-ribbed branches. Hennings made no mention of the rib condition.

Bruhn and Holmstedt (1974) succeeded in chemically analyzing old peyote material collected by Dr. J. R. Briggs and given to Sereno Watson at Harvard University in 1887. It is quite likely that specimens from this same collection were also given to Dr. Louis Lewin in Germany through Parke, Davis and Company. Hennings, in turn, used some of them to describe his new species, *Anhalonium lewinii*. The chemical studies of Bruhn and Holmstedt support the morphological data and show that this material belongs to *L. williamsii*. The Hennings taxon, *A. lewinii*, therefore must be considered as a synonym of *L. williamsii*.

Echinocactus williamsii var. *lutea* Rouhier, Trav. Labouret Mat. Med. and Pharm. 17, chap 5: 65. 1926, *Lophophora echinata* var. *lutea* Croizat, Des. Pl. Life 16: 44. 1944, *Lophophora williamsii* var. *lutea* Soulaire, Cact. et Med. 121. 1947, and *Lophophora lutea* Backeb., Die Cactaceae 5: 2901. 1961, were all based on the observation by Rouhier that the appearance of the tubercles varied somewhat from that of the "typical" peyote and that the flowers supposedly were yellow. Rouhier did not designate a type specimen, nor have yellow flowers been observed in any distinct population of peyote.

Lophophora echinata Croizat, Des. Pl. Life 16: 43. 1944 and *Lophophora williamsii* var. *echinata* H. Brav.-Holl., Cact. Succ. Mex. 12(1): 12. 1967, were based on a photograph by Schultes in Cact. Succ. Jour. 12: 180, fig. 3. 1940. Croizat's criterion for establishing the new species was that "*L. williamsii* can hardly be made to cover, indeed, the peculiar form with loose tubercles.... The tubercles that are flattened out, and run into definite ribs in *L. williamsii*, are endowed with distinct individuality." A relatively minor rib variation is insufficient to establish a separate species or variety because plants in single clones and populations vary too widely.

Lophophora williamsii var. *decipiens* Croizat, Des. Pl. Life 16: 44. 1944, was based on an illustration in Britton and Rose, The Cactaceae 3: pl. 10, fig. 4. 1923. Croizat stated that the vegetative body of this variety was basally tubercled or with distinct podaria rather than ribs, and that the flower extended out of the top of the plant to a greater extent. He designated the type to be the illustration cited above. I do not believe these characters are consistent enough in occurrence to justify separate taxonomic status.

Lophophora williamsii var. *pentagona* Croizat, Des. Pl. Life 16: 44. 1944, was based on a photograph by Y. Wright in Cact. Succ. Jour. 3: 55. 1931, titled "Anhalonium sp. undetermined." It is an illustration of a young 5-ribbed plant which commonly occurs in populations having plants with branches bearing 7–14 ribs. Rib number alone is insufficient basis for establishing a separate taxon.

Lophophora williamsii var. *pluricostata* Croizat, Des. Pl. Life 16: 9. 1944, was proposed because Croizat believed specimens of peyote having 13 ribs and which formed clusters were worthy of separate taxonomic rank. He offered as a type an illustration by Schultes in Cact. Succ. Jour. 12: 178, fig. 1. 1940. As stated earlier, rib number is highly variable depending on the age and health of the plant; moreover, clusters of heads are also common, especially if the plants have been injured.

Echinocactus williamsii, "Hylaeid," β *anhaloninica* (= *E. lewinii*) Schumann in Engler, Bot. Jahrb. 24: 567. 1898, nom. provisorum, and *Echinocactus williamsii* var. *anhaloninica* Rouhier, Trav. Labouret Mat. Med. and Pharm. 17, chap. 5: 70. 1926. Schumann apparently believed that, although there was but a single species of peyote, there were two distinct chemical forms as shown by the studies of Lewin, Heffter, and other European chemists. This epithet should be considered as a synonym of *L. williamsii*.

Lophophora fricii Haberm., Kaktusy 10 (6): 123–27, 144. 1974, is described as differing from *L. williamsii* in having gray-green epidermis, a different arrangement of ribs, "carmine-red flowers," and seeds with a coarse testa and a "compressed V-shaped hilum." The plant upon which the description was based was said to be from the vicinity of San Pedro, Coahuila, Mexico, and was collected by Dennis Cowper. The holotype is reported as specimen L-5 in the herbarium at the Department of Biology, Biophysics and Biochemistry of the Charles University Medical Facility in Plzen, Czechoslovakia, in a later article in English (Habermann 1975). In the *Cactus and Succulent Journal* (U.S.) Habermann also amplified his description of *L. fricii*, particularly emphasizing its large size (up to 40 cm diam) and great number of ribs (13–21). Neither of these supposedly significant characters was so described in the validating Latin description of 1974. Until it can be substantiated by evidence that there is a natural population of *Lophophora* in Coahuila that consistently possesses a unique set of morphological features such as Habermann has described, it appears appropriate to conclude that the proposed taxon does not represent a distinct population but rather is a somewhat unusual cultivated individual of the widespread and variable species *L. williamsii*.

Lophophora jourdaniana Haberm., Kaktusy 11 (1): 3–6, 24. 1975, is described as differing from any other species of *Lophophora* by having a rose-violet perianth, pistil, and filaments; the presence of small and persistent spines on young areoles; and cleistogamic flowers. Apparently this latter term is misunderstood or used incorrectly, for the photograph accompanying the type description shows a plant with an open flower; Habermann comments that "it is not possible to fertilize them with the pollen of Loph. williamsii, varieties of her, nor with the pollen of Loph. diffusa and Loph. fricii." In a later article in English, Habermann (1975) stated that the flowers are not cleistogamic but persisted in emphasizing the lack of pollen compatibility. He further commented in the American journal that "the description was based on a specimen purchased from K. H. Uhlig as a fresh import from Mexico," but that the exact locality was unknown. The holotype was stated to be L-2 deposited in the herbarium at the Department of

Biology, Biophysics and Biochemistry of the Charles University Medical Facility in Plzen, Czechoslovakia.

There is no indication that the morphological characters which distinguish this proposed taxon are typical and consistent for a natural population of *Lophophora*; it seems more likely that the Uhlig specimen is a somewhat unusual variant individual of *L. williamsii*.

Habermann, in describing this new taxon, chose a specific epithet that has long been used horticulturally in connection with *Lophophora*; unfortunately, prior to 1975 the name had never been validly published. It has appeared in the following ways: *Anhalonium jourdanianum* Lewin, Ber. Deutsch. Bot. Ges. 12: 289. 1894, nom. nud., *Echinocactus jourdanianus* Rebut ex Maas, Monat. Kakteenk. 15: 122. 1905, nom. nud., and *Lophophora jourdaniana* Kreuz., Verzeichnis 9. 1935, as synonym of *A. jourdanianum* Lewin. There is no clear indication to what form of peyote these names referred; however, Habermann (1975) stated that in recent years the epithet has been applied to specimens having violet-red flowers. It is unfortunate that he chose to legitimize an epithet that was never validly published and for which there has been no way to determine the original purpose of its use. Therefore, it is impossible even to list the earlier binomials which utilize the epithet *jourdanianum* as synonyms—or even indicating the same type of plant—of the new Habermann species.

More recent chemical studies of the proposed species, *L. fricii* and *L. jourdanianum*, by Habermann (1977; 1978) do not clarify the status of these taxa, especially when *L. fricii* from Coahuila and within the center of distribution of *L. williamsii* is reported to have an alkaloid constitution similar to that of *L. diffusa* from Querétaro.

2. *L. DIFFUSA* (Croizat) H. Brav.-Holl., Cact. Succ. Mex. 12: 13. 1967.

TYPE: Fig. 201 in Bravo, *Las Cactaceas de México* 378. 1937. Bravo, when elevating the epithet to the specific level, stated that the plant in the photograph was collected in 1935 by Carolina Schmoll near Vizarrón, Querétaro.

Plants soft, yellow-green, often somewhat globular in shape, 2–7 cm high, 5–12 cm in diam; ribs usually absent, podaria rarely elevated but broad and flat; areoles 1–2 cm apart, 2–3 mm in diam; flower 1.3–2.2 cm in diam, 1.3–2.4 cm long; outer perianth segments 1–2 mm broad; inner perianth segments 2–2.5 mm broad, usually white or faintly pink but sometimes appearing yellowish-white; pollen 0–6 colpate, 26.1–48.5 mm in diam (fig. A.2).

Occurring in Mexico within a fairly restricted range near Vizarrón in the state of Querétaro (see fig. 8.2).

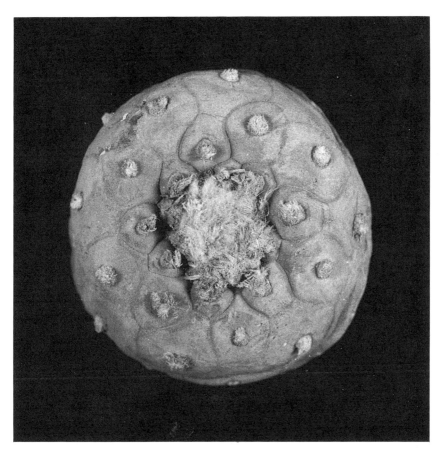

FIGURE A.2. *Lophophora diffusa* (Croizat) H. Brav.-Holl.

Representative specimens:
MEXICO: QUERETARO: without locality, *Altamirano s.n.* (US); without locality, *Willi Wagner s.n.* (POM); Mun. Cadereyta, Vizarrón, *L. Scheinvar 5594* (MEXU); N of Cadereyta, *E.F. Anderson 1656, 1659, 2300* (POM); Carretera federal de Vizarrón a Higuerillas, *L. Scheinvar 4630* (MEXU); Peña Blanca, Mpio. Peñamiller, *L. Scheinvar 2376, 3710, 3717* (MEXU).

Lophophora echinata var. *diffusa* Croizat, Des. Pl. Life 16: 44. 1944 and *Lophophora diffusa* H. Brav.-Holl., Cact. Succ. Mex. 12: 13. 1967, are both based on the Schmoll photograph of a plant collected near Vizarrón, Querétaro. The Querétaro population is isolated, self-perpetuating, and has several distinct characteristics as just described. It is deserving of recognition as a separate taxon; therefore, the illustration designated as a type by Croizat must serve as the type.

Bruhn and Holmstedt (1974) extensively discuss early reports of *L. diffusa* (or at least the population that was later given that name) in the literature. They conclude that the population was known and specimens collected as early as 1905 by F. Altamirano and Dr. J. N. Rose. Material from this population had somehow succeeded in reaching German laboratories before 1900. Who collected that material is not known.

Echinocactus williamsii "Hylaeid," α *pellotinica* (= *E. williamsii*, Type) Schumann in Engler, Bot. Jahrb. 24: 567. 1898, nom. provisorum, and *Echinocactus williamsii* var. *pellotinica* Rouhier, Trav. Labouret Mat. Med. and Pharm. 17, chap. 5: 70. 1926, have been considered as nothing more than synonyms of *E. williamsii*. This is the other chemical form of *E. williamsii* that Schumann recognized. Interestingly, the material that was the basis of this taxon almost certainly belongs to the Querétaro population of peyote now known as *L. diffusa*. It should therefore be considered a synonym of that species.

Excluded Names and Taxa of Uncertain Status

Anhalonium rungei Hildm. ex Arendt in Monat. Kakteenk. 3: 68. 1893, is a nomen nudum although it was applied occasionally to peyote by horticulturists. The name is not used in the trade today. *Anhalonium subnodusum* Hildm. ex Arendt in Monat. Kakteenk. 3: 68. 1893, should be treated in the same way.

A. D. Houghton proposed the epithet *cristata* in Cact. Succ. Jour. 2: 490. 1931. From time to time abnormal plants appear within natural populations. These cristate specimens are prized by horticulturists and are often in collections, but they are aberrant forms of the typical plant and do not deserve formal taxonomic status.

Lophophora caespitosa Fric ex Roeder, Kakteenk. 1937: 190, nom. nud., and *Lophophora williamsii* var. *caespitosa* Y. Ito, Cacti 96. 1952, are epithets which apparently refer to forms of *Lophophora* with several branches arising from the same root, a common phenomenon in natural populations.

Lophophora texana Fric ex Roeder, Kakteenk. 1937: 190, nom. nud., and *Lophophora williamsii* var. *texana* Kreuzinger ex Backeberg, Die Cactaceae 5: 2895. 1961, are without description or illustration. The epithet probably was proposed by European horticulturists for peyote plants arriving from Texas. Apparently their main source of plants was Mexico, and this may have been to differentiate the two geographical groups. Backeberg stated that Kreuzinger had used the epithet at the varietal level in his *Verzeichnis*. However, in actuality he used it at the specific level.

Lophophora ziegleri Werderm. ex Borg, Cacti 210. 1937, *Lophophora ziegleriana* Soulaire, Cact. et Med. 121. 1947, as syn., and *Lophophora tiegleri* Soulaire, Cact. et Med. 121. 1947, as syn., commonly have been known in the cactus trade as peyote plants having quite distinct, deep ribs with well-pronounced tufts of trichomes. However, Borg described them as having less depressed stems, low tubercles, smaller tufts of wool, and pale yellow flowers. None of the names were published with illustrations or descriptions and they probably refer to varying forms of L. williamsii reaching European markets.

Lophophora albilanata F. Schmoll, Katalog 2. 1947, nom. nud. and *Lophophora ziegleri* var. *albilanata* Soulaire, Cact. et Med. 121. 1947, nom. nud. apparently were epithets applied to some horticultural forms having very white trichomes.

Lophophora ziegleri var. *diagnonalis* F. Schmoll, Katalog 2. 1947, nom. nud. possibly was proposed to describe the shape or arrangement of tubercles and ribs.

Lophophora flavilanata F. Schmoll, Katalog 2. 1947, nom. nud. and *Lophophora ziegleri* var. *flavilanata* Soulaire, Cact. et Med. 121. 1947, nom. nud. may have been epithets referring to forms of peyote having golden-yellow flowers or trichomes.

Lophophora ziegleri var. *mammillaris* F. Schmoll, Katalog 2. 1947, nom. nud. probably was proposed because certain plants had protuberant podaria.

Rouhier, in his extensive treatment of peyote in 1927, presented some nomenclatural combinations that have led to considerable confusion among taxonomists. One epithet, *"pseudo-Lewinii hortulorum"* appeared on page 61 while on page 62 he referred to *"Echinocactus pseudo-Lewinii Thompsonii."* Although his discussion is unclear, it appears that he probably did not intend to create any new varieties or binomials; rather, he simply reverted to Latin when referring to certain descriptive phrases when dealing with various forms often recognized in the horticultural trade. Should these have to be accepted as legitimate epithets, they would be referred to the synonymy of *L. williamsii*.

The binomial *Anhalonium visnagra* Schumann, Monat. Kakteenk. 6: 174. 1896, nom. nud., occasionally has been referred to *Lophophora*. However, Schumann made reference to this plant as a new species in the trade along with *Echinocactus williamsii* and *E. lewinii*. There is no indication that he thought they were related.

Mammillaria cirrhifera Martius, Nov. Act. Physico-med. Acad. 16: pars 1. 1832, has also been mentioned in *Lophophora* synonymy. Martius's description obviously is not that of *Lophophora* and presently the epithet is considered to be a synonym of *Mammillaria compressa*.

Appendix B: Peyote Alkaloids

This appendix contains all of the presently known, naturally occurring alkaloids, alkaloidal amines, and amino acids of peyote (*Lophophora* spp.). Each is listed according to its currently accepted name, with probable structure, molecular formula, and melting or boiling point (if known).

For additional and more extensive discussion of the chemical nature of these substances, see the seminal papers by G. J. Kapadia and M.B.E. Fayez (1970; 1973).

MONO-OXYGENATED
PHENETHYLAMINES

1. Tyramine
   $C_8H_{11}ON$
   mp=161°

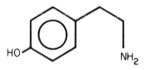

2. N-Methyltyramine
   $C_9H_{13}ON$
   mp=127–128°

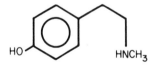

3. Hordenine
   $C_{10}H_{15}ON$
   mp=117–118°

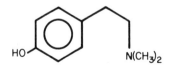

4. Candicine
   $C_{11}H_{19}O_2N$
   mp=230–231°
   (as iodide)

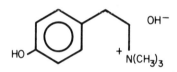

DIOXYGENATED
PHENETHYLAMINES

5. Dopamine
   $C_8H_{11}O_2N$
   mp=241°
   (as HCl)

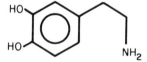

6. Epinine
 $C_9H_{13}O_2N$
 mp=188–189°

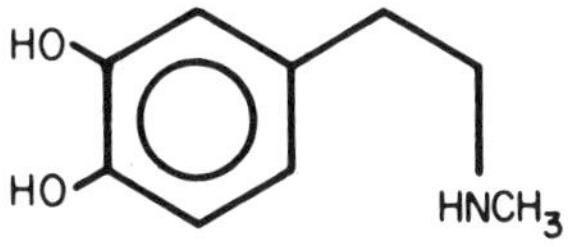

7. 4-Hydroxy-3-methoxy-
 phenethylamine
 $C_9H_{13}O_2N$
 ?

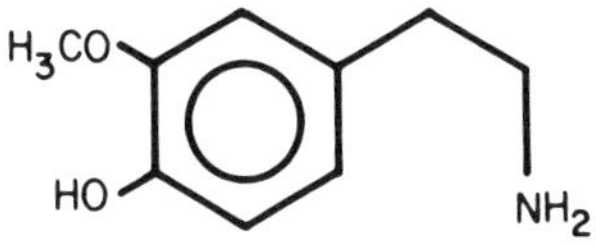

8. N-Methyl-4-hydroxy-3
 methoxyphenethylamine
 $C_{10}H_{15}O_2N$
 mp=154–155°
 (as HCl)

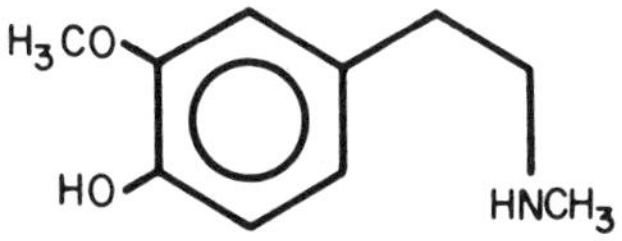

9. N,N-Dimethyl-4-hydroxy-3
 methoxyphenethylamine
 $C_1H_{17}O_2N$
 mp=190–191°
 (as HCl)

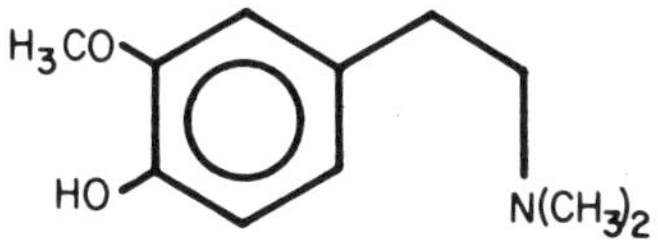

10. 3,4-Dimethoxyphenethylamine
 $C_{10}H_{15}O_2N$
 bp=188°/
 15 mm

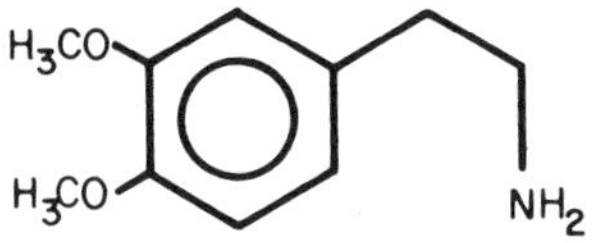

TRIOXYGENATED PHENETHYLA-
MINES AND RELATED AMIDES

11. 3,4-Dihydroxy-5-methoxy-
 phenethylamine
 $C_9H_{13}O_3N$
 mp=207°

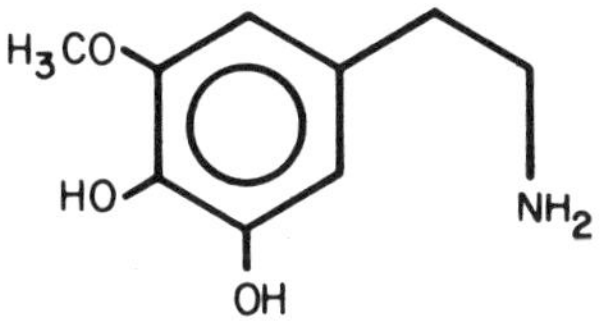

12. 3-Hydroxy-4,5-dimethoxy-
 phenethylamine
 (3-Demethylmescaline)
 $C_{10}H_{15}O_3N$
 mp=178–179°
 (as HCl)

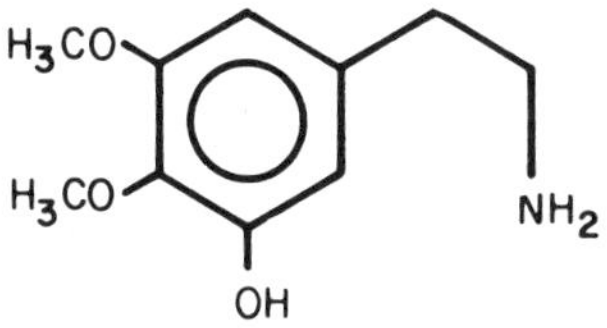

Peyote Alkaloids : 221

13. N-Methyl-3-hydroxy-4,
5-dimethoxyphenethylamine
$C_{11}H_{17}O_3N$
mp=151–155°
(as HCl)

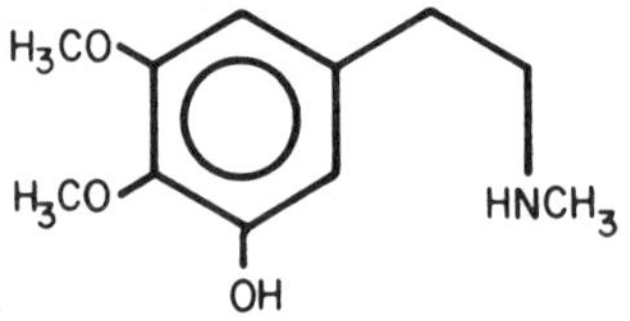

14. N,N-Dimethyl-3-hydroxy-4,
5-dimethoxyphenethylamine
$C_{12}H_{19}O_3N$
mp=180–185°
(as HCl)

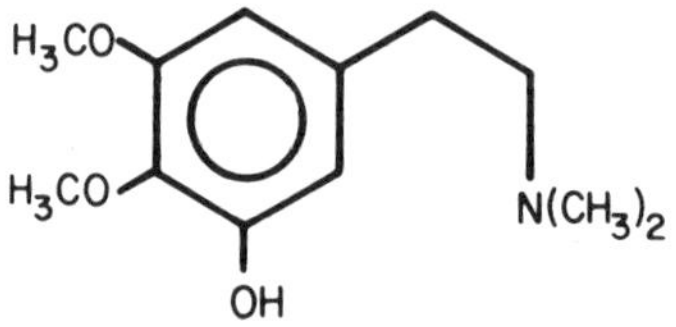

15. N-Formyl-3-hydroxy-4,
5-dimethoxyphene-thylamine
(N-Formyl-3-demethylmescaline)
$C_{11}H_{15}O_4N$
?

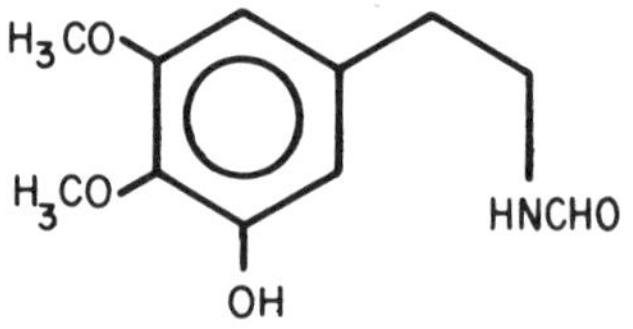

16. N-Acetyl-3-hydroxy-4,
5-dimethoxyphene-thylamine
(N-Acetyl-3-demethylmescaline)
$C_{12}H_{17}O_4N$
mp=102–103°

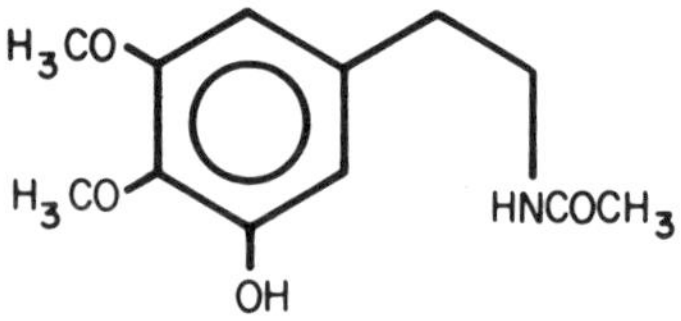

17. Mescaline (3,4,5-Trimethoxy-beta-phenethylamine)
$C_{11}H_{17}O_3N$
mp=30–32°
bp=183–186°

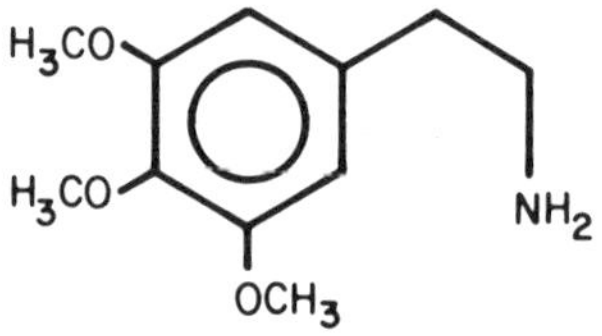

18. N-Methylmescaline
$C_{12}H_{19}O_3N$
mp=177.5–178°
(as picrate)

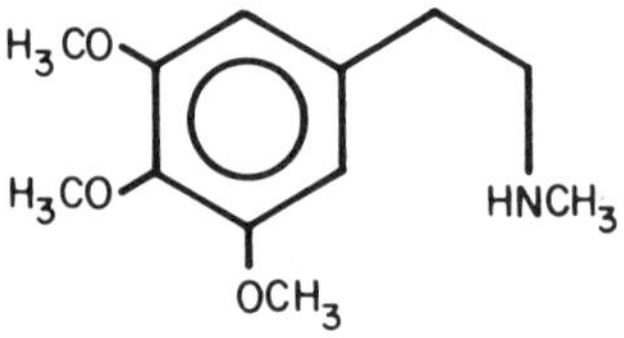

19. N-Formylmescaline
    $C_{12}H_{17}O_4N$
    mp=68–69°

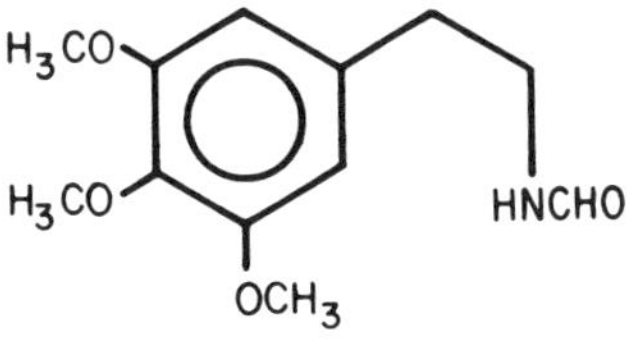

20. N-Acetylmescaline
    $C_{13}H_{19}O_4N$
    mp=93–94°

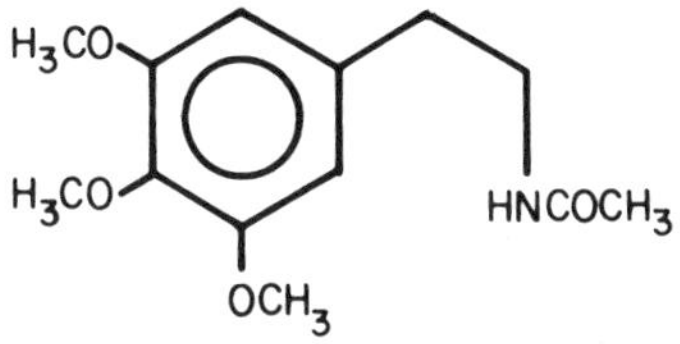

TETRAHYDRO-
ISOQUINOLINES
AND RELATED AMIDES

21. Anhalamine
    $C_{11}H_{15}O_3N$
    mp=189–191°

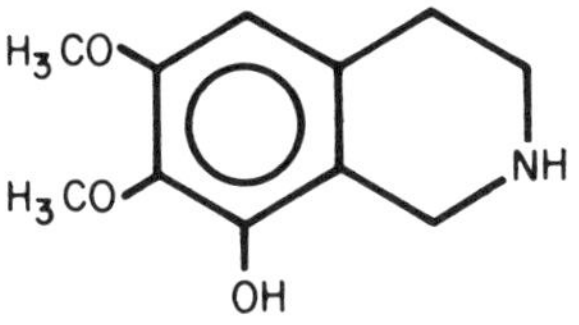

22. N-Formylanhalamine
    $C_{12}H_{15}O_4N$
    ?

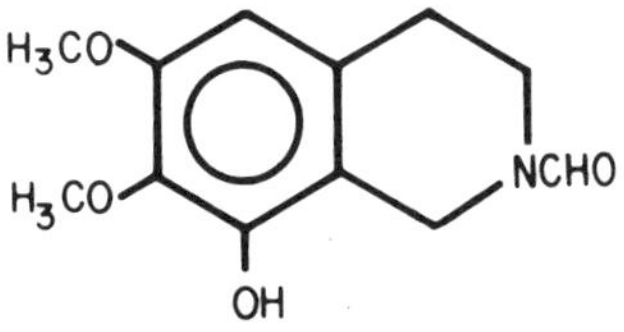

23. N-Acetylanhalamine
    $C_{13}H_{17}O_4N$
    ?

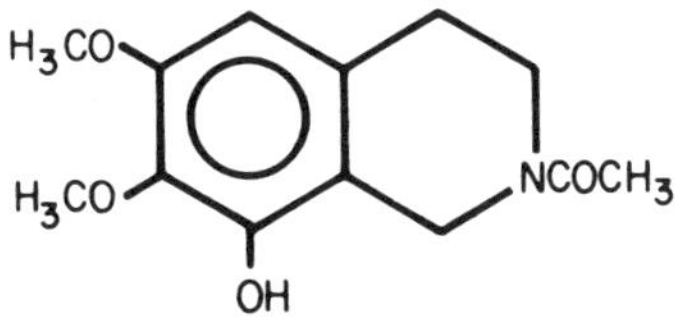

24. Isoanhalamine
    $C_{11}H_{15}O_3N$
    mp=213–215°
    (as HBr)

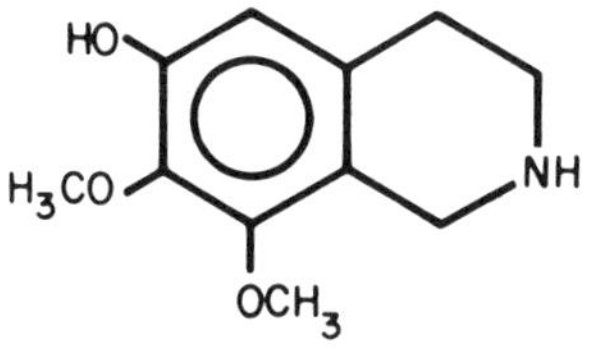

25. Anhalinine
    $C_{12}H_{17}O_3N$
    mp=61–63°

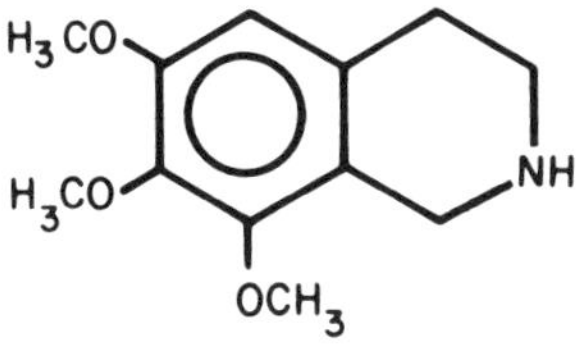

26. N-Formylanhalinine
    $C_{13}H_{17}O_4N$
    ?

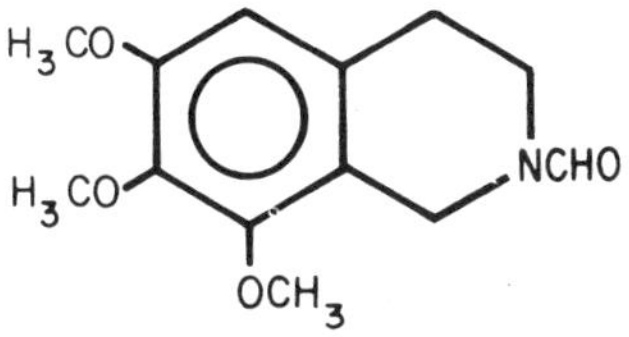

Peyote Alkaloids : 223

27. Anhalidine
    $C_{12}H_{17}O_3N$
    mp=131–133°

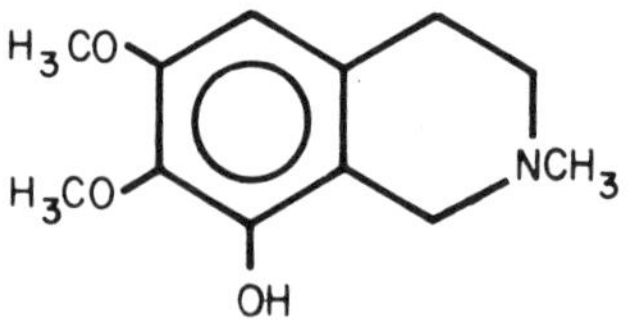

28. Anhalotine (as an iodide)
    $C_{13}H_{20}O_3N$
    mp=219–220°

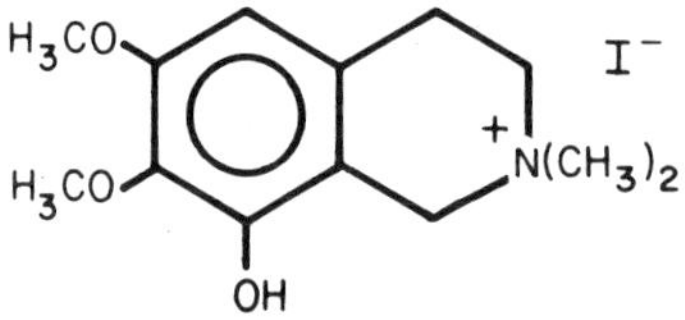

29. Isoanhalidine
    $C_{12}H_{17}O_3N$
    mp=215–218°
    (as HCl)

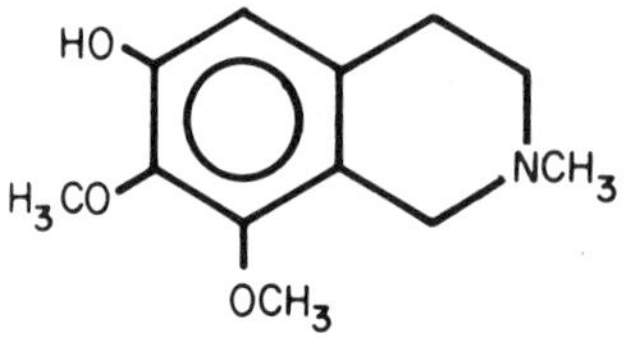

30. Anhalonidine
    $C_{12}H_{17}O_3N$
    mp=160–161°

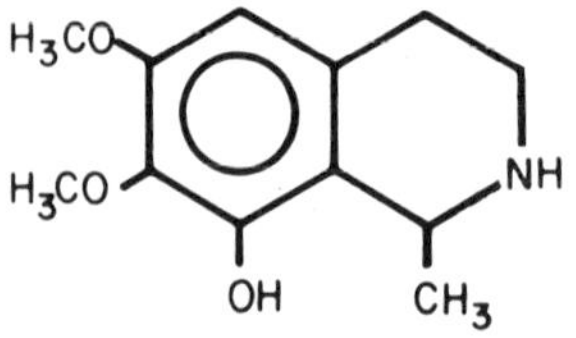

31. N-Formylanhalonidine
    $C_{13}H_{17}O_4N$
    ?

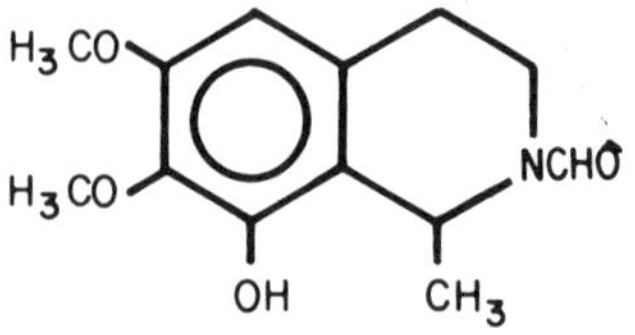

32. Isoanhalonidine
    $C_{12}H_{17}O_3N$
    mp=209–211°
    (as HBr)

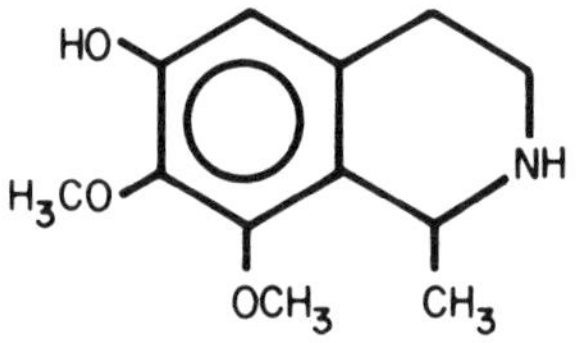

33. S-(+)-O-Methylanhalonidine
    $C_{13}H_{19}O_3N$
    bp=140°

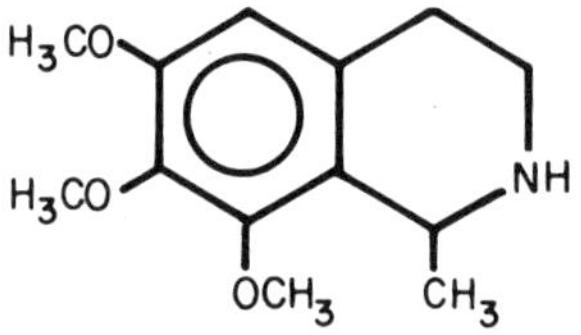

34. N-Formyl-O-methylanhalonidine
    $C_{14}H_{19}O_4N$
    ?

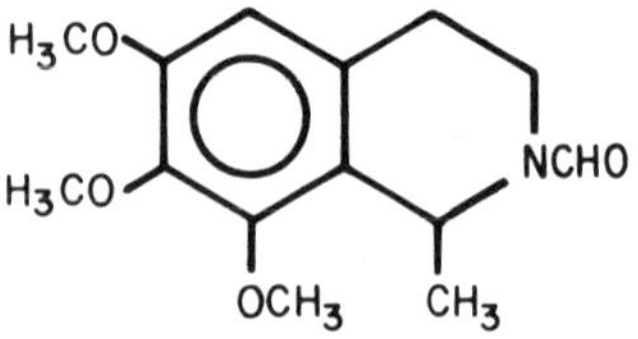

35. Pellotine
   $C_{13}H_{19}O_3N$
   mp=111–112°

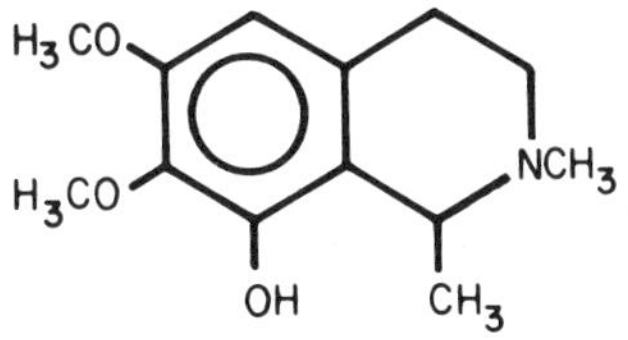

36. O-Methylpellotine
   $C_{14}H_{21}O_3N$
   ?

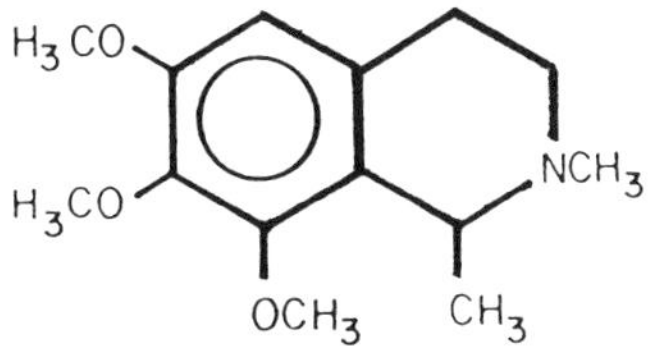

37. Peyotine (as iodide)
   $C_{14}H_{22}O_3NI$
   mp=185–186°

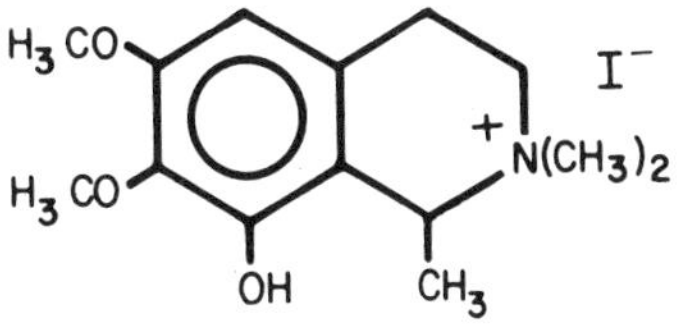

38. Isopellotine
   $C_{13}H_{19}O_3N$
   mp=212–222°
   (as HCl)

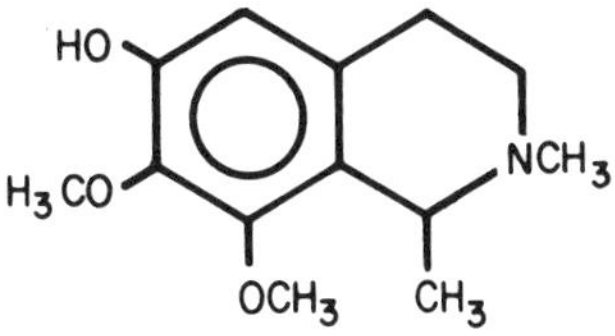

39. S-(-)-Anhalonine
   $C_{12}H_{15}O_3N$
   mp=85.5°

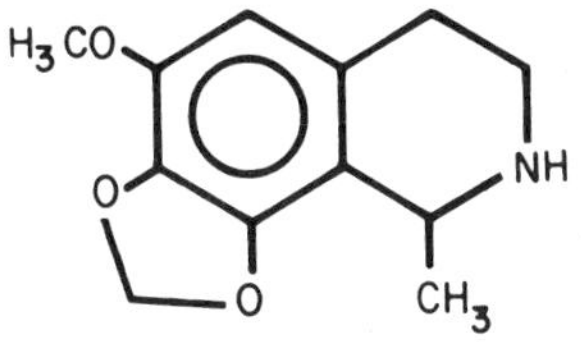

40. N-Formylanhalonine
   $C_{13}H_{15}O_4N$
   ?

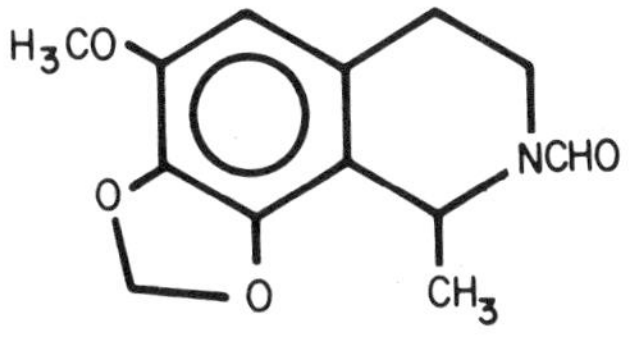

41. N-Acetylanhalonine
   $C_{14}H_{17}O_4N$
   ?

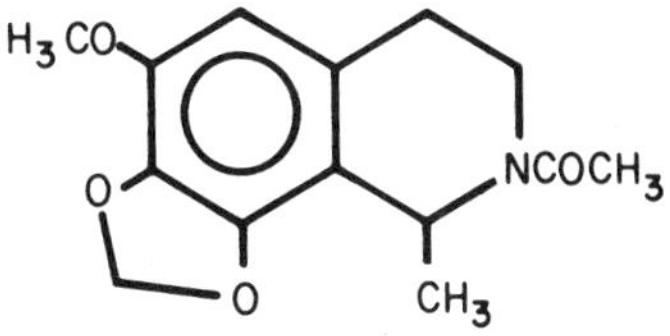

42. S-(-)-Lophophorine
   $C_{13}H_{17}O_3N$
   mp=−47°
   (CHCl$_3$)

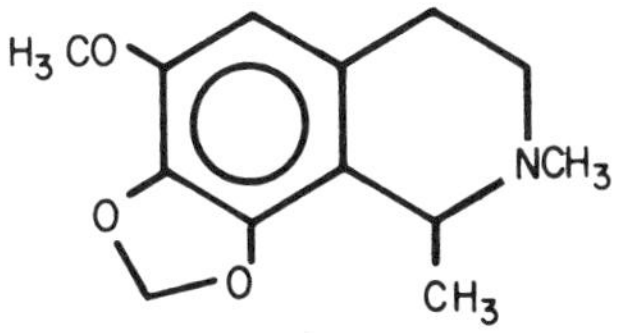

Peyote Alkaloids : 225

43. Lophotine (as iodide)
    $C_{14}H_{20}O_3N$
    mp=240–242°

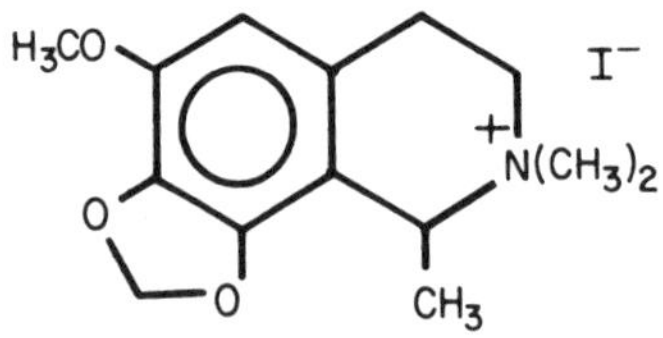

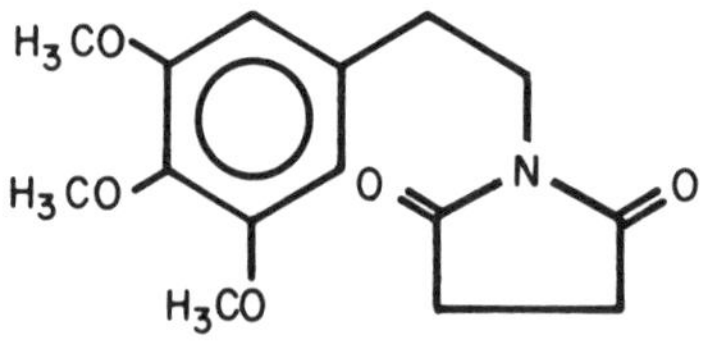

46. Mescaline malimide
    $C_{15}H_{19}O_6N$
    ?

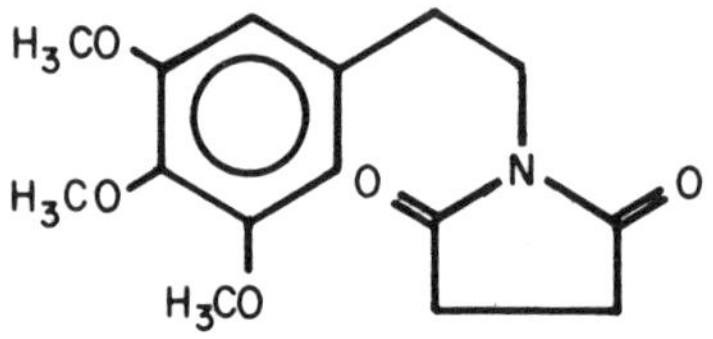

44. Peyophorine
    $C_{14}H_{19}O_3N$
    mp=155–156°
    (as picrate)

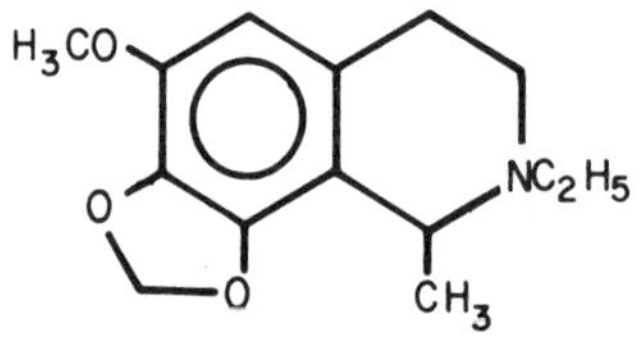

47. Mescaline citrimide
    $C_{17}H_{21}O_8N$
    ?

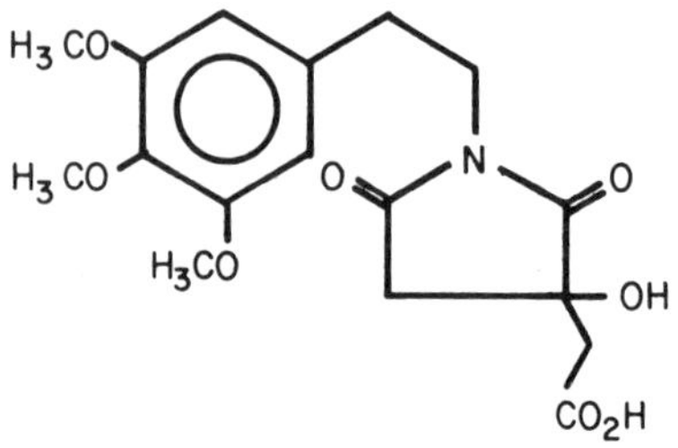

CONJUGATES WITH KREBS ACIDS

48. Mescaline maleimide
    $C_{15}H_{17}O_5N$
    ?

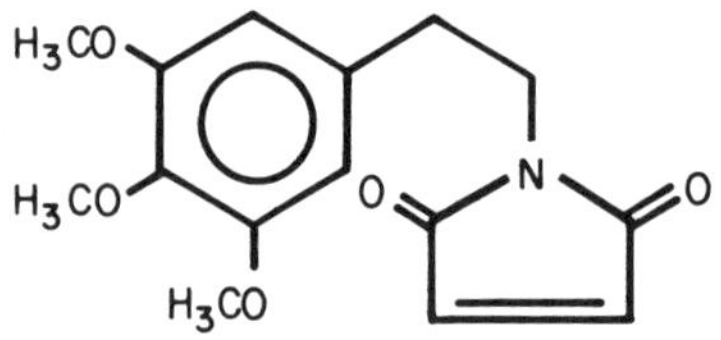

45. Mescaline succinimide
    $C_{15}H_{19}O_5N$
    mp=125–126°

226 : Appendix B

49. Mescaline isocitrimide lactone
$C_{17}H_{19}O_7N$
?

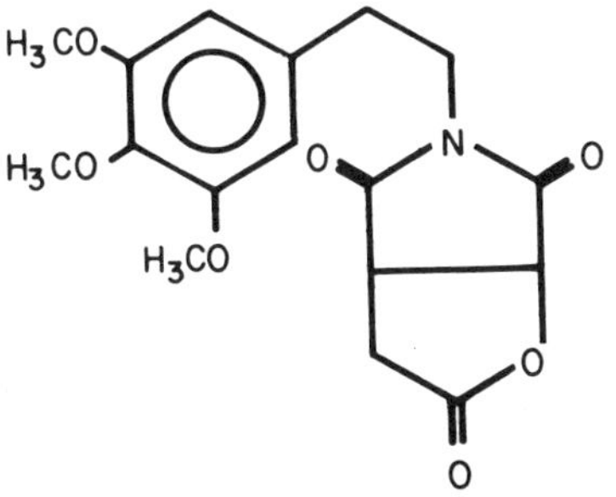

50. Peyoglutam
$C_{14}H_{17}O_4N$
mp=217–219°

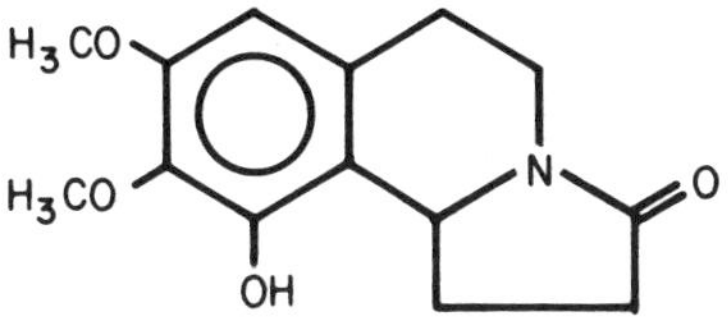

51. Mescalotam
$C_{15}H_{19}O_4N$
?

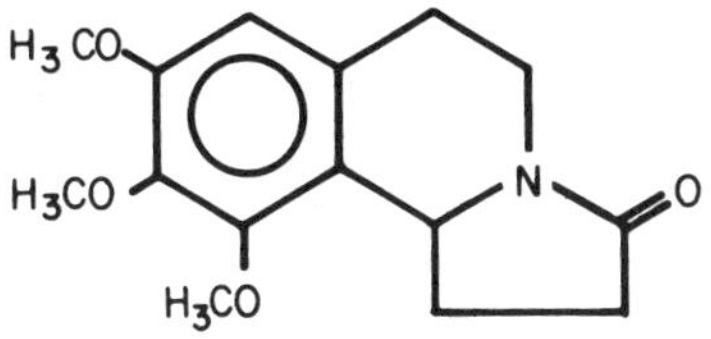

52. Mescaloxylic acid
$C_{13}H_{19}O_5N$
mp=187–189°

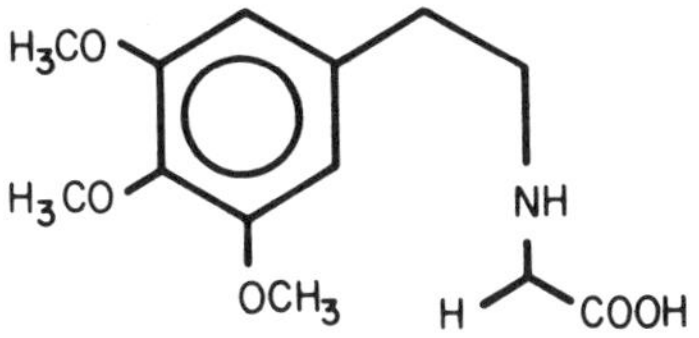

53. Mescaloruvic acid
$C_{14}H_{21}O_5N$
mp=235–236.5°

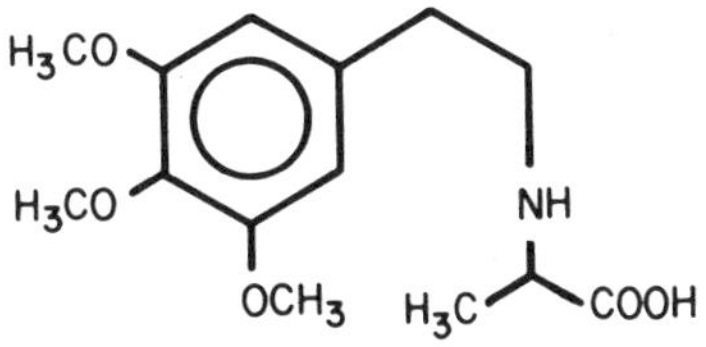

54. Peyoxylic acid
$C_{12}H_{15}O_5N$
mp=237–238°

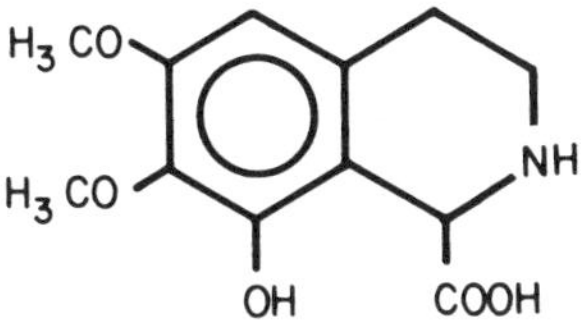

55. Peyoruvic acid
$C_{13}H_{17}O_5N$
mp=233–234°

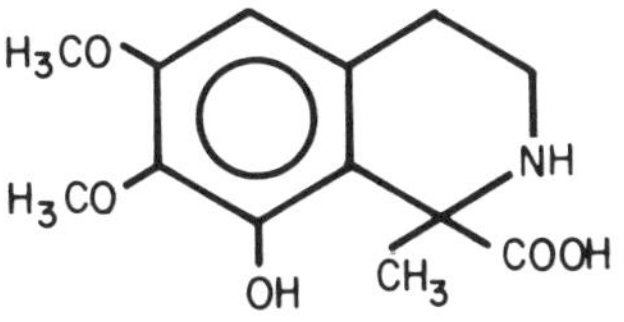

PYRROLE DERIVATIVES

56. Peyonine

$C_{16}H_{19}O_5N$
mp=131–133.5°

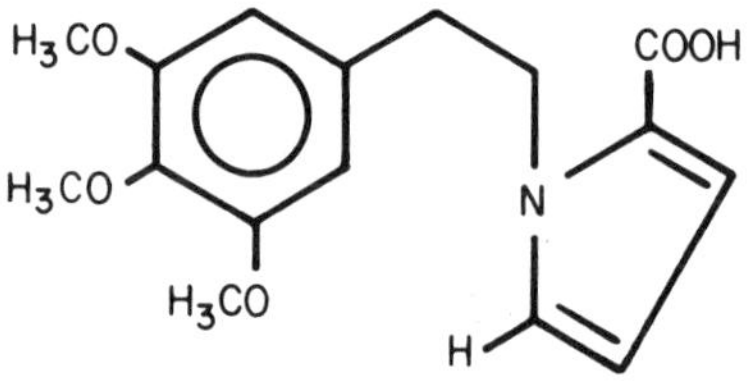

57. Peyoglunal

$C_{17}H_{21}O_5N$
?

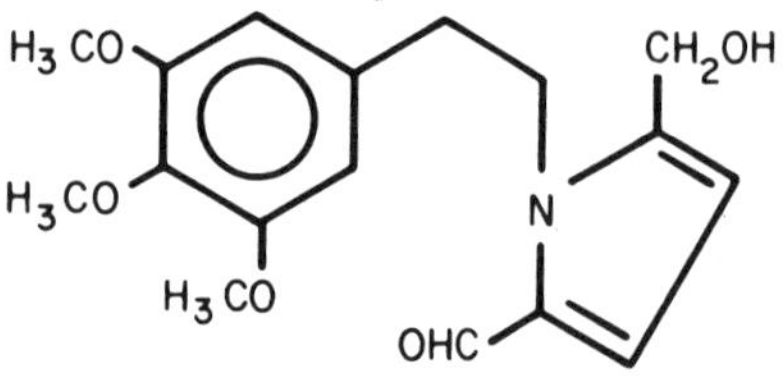

228 : *Appendix B*

Appendix C: Excerpts of Current Federal Laws and Regulations Pertaining to Peyote

The following federal laws and regulations have been passed since 1970. Those portions pertinent to peyote and its legal status at the federal level are quoted below. See chapter 9 for an extensive discussion of the legal aspects of peyote.

The Comprehensive Drug Abuse Prevention and Control Act of 1970 (Public Law 91-513)

Be it enacted by the Senate and House of Representatives of the United States of America in Congress assembled, That this Act may be cited as the "Comprehensive Drug Abuse Prevention and Control Act of 1970."...

DEFINITIONS

...

(9) The term "depressant or stimulant substance" means—

(A) a drug which contains any quantity of (i) barbituric acid or any of the salts of barbituric acid; or (ii) any derivative of barbituric acid which has been designated by the Secretary as habit forming under section 502(d) of the Federal Food, Drug, and Cosmetic Act (21 U.S.C. 352(d)); or

(B) a drug which contains any quantity of (i) amphetamine or any of its optical isomers; (ii) any salt of amphetamine or any salt of an optical isomer of amphetamine; or (iii) any substance which the Attorney General, after investigation, has found to be, and by regulation designated as, habit forming because of its stimulant effect on the central nervous system; or

(C) lysergic acid diethylamide; or

(D) any drug which contains any quantity of a substance which the Attorney General, after investigation, has found to have, and by regulation designated as having, a potential for abuse because of its depressant or stimulant effect on the central nervous system or its hallucinogenic effect....

(16) The term "narcotic drug" means any of the following, whether produced directly or indirectly by extraction from substances of vegetable origin, or independently by means of chemical synthesis, or by a combination of extraction and chemical synthesis:

(A) Opium, coca leaves, and opiates.

(B) A compound, manufacture, salt, derivative, or preparation of opium, coca leaves, or opiates.

(C) A substance (and any compound, manufacture, salt, derivative, or preparation thereof) which is chemically identical with any of the substances referred to in clause (A) or (B).

Such term does not include decocainized coca leaves or extracts of coca leaves, which extracts do not contain cocaine, or ecgonine. . . .

AUTHORITY AND CRITERIA FOR CLASSIFICATION OF SUBSTANCES

Sec. 201. (a) The Attorney General shall apply the provisions of this title to the controlled substances listed in the schedules established by section 202 of this title and to any other drug or other substance added to such schedules under this title. Except as provided in subsections (d) and (e), the Attorney General may by rule—

(1) add to such a schedule or transfer between such schedules any drug or other substance if he—

(A) finds that such drug or other substance has a potential for abuse, and

(B) makes with respect to such drug or other substance the findings prescribed by subsection (b) of section 202 for the schedule in which such drug is to be placed; or

(2) remove any drug or other substance from the schedules if he finds that the drug or other substance does not meet the requirements for inclusion in any schedule. . . .

SCHEDULES OF CONTROLLED SUBSTANCES

Sec. 202. (a) There are established five schedules of controlled substances, to be known as schedules I, II, III, IV, and V. Such schedules shall initially consist of the substances listed in this section. The schedules established by this section shall be updated and republished on a semiannual basis during the two-year period beginning one year after the date of enactment of this title and shall be updated and republished on an annual basis thereafter.

(b) Except where control is required by United States obligations under an international treaty, convention, or protocol, in effect on the effective date of this part, and except in the case of an immediate precursor, a drug or other substance may not be placed in any schedule unless the findings required for such schedule are made with respect to such drug or other substance. The findings required for each of the schedules are as follows:

(1) SCHEDULE I.—

(A) The drug or other substance has a high potential for abuse.

(B) The drug or other substance has no currently accepted medical use in treatment in the United States.

(C) There is a lack of accepted safety for use of the drug or other substance under medical supervision....

SCHEDULE I

...Hallucinogenic substances.

(c) Unless specifically excepted or unless listed in another schedule any material, compound, mixture, or preparation, which contains any quantity of the following hallucinogenic substances, or which contains any of their salts, isomers, and salts of isomers whenever the existence of such salts, isomers, and salts of isomers is possible within the specific chemical designation:

(1) 3,4-methylenedioxy amphetamine
(2) 5-methoxy-3,4-methylenedioxy amphetamine
(3) 3,4,5-trimethoxy amphetamine
(4) Bufotenine
(5) Diethyltryptamine
(6) Dimethyltryptamine
(7) 4-methyl-2,5-dimethoxyamphetamine
(8) Ibogaine
(9) Lysergic acid diethylamide
(10) Marihuana
(11) Mescaline
(12) Peyote
(13) N-ethyl-3-piperidyl benzilate
(14) N-methyl-3-piperidyl benzilate
(15) Psilocybin
(16) Psilocyn
(17) Tetrahydrocannabinols

PART D—OFFENSES AND PENALTIES

PROHIBITED ACTS A—PENALTIES

Sec. 401. (a) Except as authorized by this title, it shall be unlawful for any person knowingly or intentionally—

(1) to manufacture, distribute, or dispense, or possess with intent to manufacture, distribute, or dispense, a controlled substance; or

(2) to create, distribute, or dispense, or possess with intent to distribute or dispense, a counterfeit substance.

(b) Except as otherwise provided in section 405, any person who violates subsection (a) of this section shall be sentenced as follows:

(1) (A) In the case of a controlled substance in schedule I or II which is a narcotic drug, such person shall be sentenced to a term of imprisonment of not more than 15 years, a fine of not more than $25,000, or both. . . .

(B) In the case of a controlled substance in schedule I or II which is not a narcotic drug or in the case of any controlled substance in schedule III, such person shall be sentenced to a term of imprisonment of not more than 5 years, a fine of not more than $15,000, or both. If any person commits such a violation after one or more prior convictions of him for an offense punishable under this paragraph, or for a felony under any other provision of this title or title III or other law of the United States relating to narcotic drugs, marihuana, or depressant or stimulant substances, have become final, such person shall be sentenced to a term of imprisonment of not more than 10 years, a fine of not more than $30,000, or both. Any sentence imposing a term of imprisonment under this paragraph shall, in the absence of such a prior conviction, impose a special parole term of at least 2 years in addition to such term of imprisonment and shall, if there was such a prior conviction, impose a special parole term of at least 4 years in addition to such term of imprisonment. . . .

PENALTY FOR SIMPLE POSSESSION; CONDITIONAL DISCHARGE AND EXPUNGING OF RECORDS FOR FIRST OFFENSE

Sec. 404. (a) It shall be unlawful for any person knowingly or intentionally to possess a controlled substance unless such substance was obtained directly, or pursuant to a valid prescription or order, from a practitioner, while acting in the course of his professional practice, or except as otherwise authorized by this title or title III. Any person who violates this subsection shall be sentenced to a term of imprisonment of not more than one year, a fine of not more than $5,000, or both, except that if he commits such offense after a prior conviction or convictions under this subsection have become final, he shall be sentenced to a term of imprisonment of not more than 2 years, a fine of not more than $10,000, or both.

(b) (1) If any person who has not previously been convicted of violating subsection (a) of this section, any other provision of this title or title III, or any other law of the United States relating to narcotic drugs, marihuana, or depressant or stimulant substances, is found guilty of a violation of subsection (a) of this section after trial or upon a plea of guilty, the court may, without entering a judgment of guilty and with the consent of such person, defer further proceedings and place him on probation upon such reasonable conditions as it may require and for such period, not to exceed one year, as the court may prescribe. . . .

Special Exempt Persons (*Federal Register* 36(80): 7802. 1971)
21 CFR 1307.31 Native American Church.

The listing of peyote as a controlled substance in schedule I does not apply to the nondrug use of peyote in bona fide religious ceremonies of the Native American Church, and members of the Native American Church so using peyote are exempt from registration. Any person who manufactures peyote for or distributes peyote to the Native American Church, however, is required to obtain registration annually and to comply with all other requirements of law.

The American Indian Religious Freedom Act
(Public Law 95-341 — Aug. 11, 1978)

Whereas the freedom of religion for all people is an inherent right, fundamental to the democratic structure of the United States and is guaranteed by the First Amendment of the United States Constitution:

Whereas the United States has traditionally rejected the concept of a government denying individuals the right to practice their religion and, as a result, has benefited from a rich variety of religious heritages in this country;

Whereas the religious practices of the American Indian (as well as Native Alaskan and Hawaiian) are an integral part of their culture, tradition and heritage, such practices forming the basis of Indian identity and value systems;

Whereas the traditional American Indian religions, as an integral part of Indian life, are indispensable and irreplaceable;

Whereas the lack of a clear, comprehensive, and consistent Federal policy has often resulted in the abridgment of religious freedom for traditional American Indians;

Whereas such religious infringements result from the lack of knowledge or the insensitive and inflexible enforcement of Federal policies and regulations premised on a variety of laws;

Whereas such laws were designed for such worthwhile purposes as conservation and preservation of natural species and resources but were never intended to relate to Indian religious practices and, therefore, were passed without consideration of their effect on traditional American Indian religions;

Whereas such laws and policies often deny American Indians access to sacred sites required in their religions, including cemeteries;

Whereas such laws at times prohibit the use and possession of sacred objects necessary to the exercise of religious rites and ceremonies;

Whereas traditional American Indian ceremonies have been intruded upon, interfered with, and in a few instances banned: Now, therefore, be it

Resolved by the Senate and House of Representatives of the United States of America in Congress assembled, That henceforth it shall be the policy of the United States to protect and preserve for American Indians their inherent right of freedom to believe, express, and exercise the traditional religions of the American Indian, Eskimo, Aleut, and Native Hawaiians, including but not limited to access to sites, use and possession of sacred objects, and the freedom to worship through ceremonials and traditional rites. . . .

The Religious Freedom Restoration Act of 1993
(Public Law 103-141 — Nov. 16, 1993)

Sec. 2. CONGRESSIONAL FINDINGS AND DECLARATION OF PURPOSES.

(A) FINDINGS. — The Congress finds that —

(1) the framers of the Constitution, recognizing free exercise of religion as an unalienable right, secured its protection in the First Amendment to the Constitution;

(2) laws "neutral" toward religion may burden religious exercise as surely as laws intended to interfere with religious exercise;

(3) governments should not substantially burden religious exercise without compelling justification;

(4) in *Employment Division v. Smith* 494 U.S. 872 (1990) the Supreme Court virtually eliminated the requirement that the government justify burdens on religious exercise imposed by laws neutral toward religion; and

(5) the compelling interest test as set forth in prior Federal court rulings is a workable test for striking sensible balances between religious liberty and competing prior governmental interests.

(B) PURPOSES. — The purposes of this Act are —

(1) to restore the compelling interest test as set forth in *Sherbert v. Verner* 374 U.S. 398 (1963) and *Wisconsin v. Yoder,* 406 U.S. 205 (1972) and to guarantee its application in all cases where free exercise of religion is substantially burdened; and

(2) to provide a claim or defense to persons whose religious exercise is substantially burdened by government.

SEC. 3. FREE EXERCISE OF RELIGION PROTECTED.

(a) IN GENERAL.—Government shall not substantially burden a person's exercise of religion even if the burden results from a rule of general applicability, except as provided in subsection (b).

(b) EXCEPTION.—Government may substantially burden a person's exercise of religion only if it demonstrates that application of the burden to the person—

(1) is in furtherance of a compelling governmental interest; and

(2) is the least restrictive means of furthering that compelling governmental interest. . . .

American Indian Religious Freedom Act Amendments of 1994
(Public Law 103-344 [H.R. 4230]—Oct. 6, 1994)

An Act to amend the American Indian Religious Freedom Act to provide for the traditional use of peyote by Indians for religious purposes, and for other purposes. *Be it enacted by the Senate and House of Representatives of the United States of America in Congress assembled,*

SECTION 1. SHORT TITLE.

This Act may be cited as the "American Indian Religious Freedom Act Amendments of 1994".

SECTION 2. TRADITIONAL INDIAN RELIGIOUS USE
OF THE PEYOTE SACRAMENT.

The Act of August 11, 1978 (42 U.S.C. 1996), commonly referred to as the "American Indian Religious Freedom Act," is amended by adding at the end thereof the following new section:

"SEC. 3.(a) The Congress finds and declares that—

"(1) for many Indian people, the traditional ceremonial use of the peyote cactus as a religious sacrament has for centuries been integral to a way of life, and significant in perpetuating Indian tribes and cultures;

"(2) since 1965, this ceremonial use of peyote by Indians has been protected by Federal regulation;

"(3) while at least 28 States have enacted laws which are similar to, or are in conformance with, the Federal regulation which protects the ceremonial use of peyote by Indian religious practitioners, 22 States have not done so, and this lack of uniformity has created hardship for Indian people who participate in such religious ceremonies;

"(4) the Supreme Court of the United States, in the case of *Employment Division v. Smith*, 494 U.S. 872 (1990), held that the First Amendment does not protect Indian practitioners who use peyote in Indian religious ceremonies, and also raised uncertainty whether the religious practice would be protected under the compelling State interest standard; and

"(5) the lack of adequate and clear legal protection for the religious use of peyote by Indians may serve to stigmatize and marginalize Indian tribes and cultures, and increase the risk that they will be exposed to discriminatory treatment.

"(b)(1) Notwithstanding any other provision of law, the use, possession, or transportation of peyote by an Indian for bona fide traditional ceremonial purposes in connection with the practice of a traditional Indian religion is lawful, and shall not be prohibited by the United States or any State. No Indian shall be penalized or discriminated against on the basis of such use, possession or transportation, including, but not limited to, denial of otherwise applicable benefits under public assistance programs.

"(2) This section does not prohibit such reasonable regulation and registration of those persons who cultivate, harvest, or distribute peyote as may be consistent with the purposes of this Act. . . .

"(7) (c) For the purposes of this section—

"(1) the term 'Indian' means a member of an Indian tribe;

"(2) the term 'Indian tribe' means any tribe, band, nation, pueblo, or other organized group or community of Indians, including Aslaska Native village (as defined in, or established pursuant to, the Alaska Native Claims Settlement Act [43 U.S.C. 1601 et seq.]), which is recognized as eligible for the special programs and services provided by the United States to Indians because of their status as Indians;

"(3) the term 'Indian religion' means any religion—

"(A) which is practiced by Indians, and

"(B) the origin and interpretation of which is from within a traditional Indian culture or community; and

"(4) the term "state" means any State of the United States, and any political subdivision thereof."

References Cited

Aberle, D. F.
    1991   *The Peyote Religion Among the Navajo.* 2d ed., reprint. Norman: University of Oklahoma Press.

Adovasio, J. M., and G. F. Fry
    1976   Prehistoric Psychotropic Drug Use in Northeastern Mexico and Trans-Pecos Texas. *Economic Botany* 30:94–96.

Aghajanian, G. K.
    1982   Neurophysiologic Properties of Psychotomimetics. In *Handbook of Experimental Pharmacology,* vol. 55/pt. 3, 89–109, ed. F. Hoffmeister and G. Stille. New York: Springer Verlag.

Agurell, S.
    1969   Cactaceae Alkaloids. I. *Lloydia* 32:206–16.

Agurell, S., J. G. Bruhn, J. Lundstrom, and U. Svensson
    1971   Cactaceae Alkaloids. X. Alkaloids of *Trichocereus* Species and Some Other Cacti. *Lloydia* 34:183–87.

de Alarcón, H. R.
    1898   Tratado de las supersticiones y costumbres gentilicas. *Anales del Museo Nacional de México* 6:123–223.

Aldana E., G.
    1971   Mesa del Nayar's Strange Holy Week. *National Geographic* 139:780–95.

Alegre, F. J.
    1958   *Historia de la Provincia de la Compañía de Jesús de Nueva España* (1841–1842). Rome: Institutum Historicum S. J.

de Alva, B.
    1634   *Confessionario mayor y menor en lengua Mexicana.* Mexico City: Francisco Salbago.

American Psychiatric Association
    1980   *Diagnostic and Statistical Manual of Mental Disorders.* 3d ed. (DSM-III). Washington, D.C.: American Psychiatric Association.

Anderson, E. F.
    1963   A Revision of *Ariocarpus* (Cactaceae). III. Formal Taxonomy of the Subgenus *Roseocactus. American Journal of Botany* 50:724–32.
    1964   A Revision of *Ariocarpus* (Cactaceae). IV. Formal Taxonomy of the Subgenus *Ariocarpus. American Journal of Botany* 51:144–51.

  1969 The Biogeography, Ecology, and Taxonomy of *Lophophora* (Cactaceae). *Brittonia* 21:299–310.
  1995 The "Peyote Gardens" of South Texas: A Conservation Crisis? *Cactus and Succulent Journal* (U.S.) 67:67–73.

Anderson, E. F., and M. S. Stone
  1971 A Pollen Analysis of *Lophophora* (Cactaceae). *Cactus and Succulent Journal* (U.S.) 43:77–82.

Appel, J. B., and D. X. Freedman
  1968 Tolerance and Cross-Tolerance Among Psychotomimetic Drugs. *Psychopharmacologia* 13:267–74.

de Arlegui, P.
  1851 *Cronica de la Provincia de N. S. P. S. S. Francisco de Zacatecas.* 2d ed. Mexico City: Cumplido.

Arth, M. J.
  1956 A Functional View of Peyotism in Omaha Culture. *Plains Anthropologist* 7:25–29.

Arthur, W. R.
  1947 *Law of Drugs and Druggists.* 3d ed. St. Paul, Minn.: West Publishing Co.

Barron, F., M. E. Jarvik, and S. Bunnell, Jr.
  1964 The Hallucinogenic Drugs. *Scientific American* 210:29–37.

Baselt, R. C., and R. H. Cravey
  1989 *Disposition of Toxic Drugs and Chemicals in Man.* 3d ed. Chicago: Chicago Yearbook Medical Publishers.

Bates, M.
  1967 Man the Drug Taker. *Natural History* 76:8–16.

Bauml, J. A., G. Voss, and P. Collings
  1990 'Uxa Identified. *Journal of Ethnobiology* 10:99–101.

Benedict, R. F.
  1922 The Vision in Plains Cultures. *American Anthropologist* 24:1–23.

Bennett, W. C., and R. M. Zingg
  1935 *The Tarahumara: An Indian Tribe of Northern Mexico.* Chicago: University of Chicago Press.

Bergman, R. L.
  1971 Navajo Peyote Use: Its Apparent Safety. *American Journal of Psychiatry* 128:695–99.

Biel, J. H., B. Bopp, and B. D. Mitchell
  1978 Chemistry and Structure-Activity Relationships of Psychotropic Drugs. Part 2. Structure-Activity Relationships. In *Principles of*

  Psychopharmacology, ed. W. G. Clark and J. del Giudice, 140–68. New York: Academic Press.

Biskupic, J.
 1991 Abortion Dispute Entangles Religious Freedom Bill. *Congressional Quarterly Weekly Report* 49:913–18.

Bittle, W. E.
 1960 The Curative Aspects of Peyotism. *Bios* 31:140–48.

Bliss, E. L., and L. D. Clark
 1962 Visual Hallucinations. In *Hallucinations*, ed. L. J. West, 92–107. New York: Grune and Stratton.

Blofeld, J.
 1966 A High Yogic Experience Achieved with Mescaline. *Psychedelic Review* 7:27–32.

Boke, N. H., and E. F. Anderson
 1970 Structure, Development, and Taxonomy in the Genus *Lophophora*. *American Journal of Botany* 57:569–78.

Bonnin, G., and R. T. Bonnin
 1937 Letter of 12 October 1916. In *Survey of Conditions of the Indians in the United States*. Hearings before a Subcommittee of the Committee on Indian Affairs, U.S. Senate, 75th Cong., 1st sess., 18296–97. Washington, D.C.: U.S. Government Printing Office.

Bourguignon, E.
 1977 Altered States of Consciousness, Myths, and Rituals. In *Drugs, Rituals, and Altered States of Consciousness*, ed. B. M. DuToit, 7–23. Rotterdam: A. A. Balkema.

Braden, W.
 1967 *The Private Sea*. Chicago: Quadrangle Books.

Bridger, W. H., I. J. Mandel, and D. M. Stoff
 1973 Mescaline: No Tolerance to Excitatory Effects. *Biological Psychiatry* 7:129–38.

Brinton, D. G.
 1894 Nagualism. *Proceedings of the American Philosophical Society* 33:11–73.

Brito, S. J.
 1989 *The Way of a Peyote Roadman*. American University Studies Series 21, Regional Studies, Vol. 1. New York: Peter Lang.

Brossi, A., F. Schenker, and W. Leimgruber
 1964 232/ Synthesen in der Isochinolinreihe neue Synthesen der

  Kaktusalkaloide Anhalamin, Anhalidin, rac. Anhalonidin, und rac. Pellotin. *Helvitica Chimica Acta* 47:2089–98.

Browne, R. G., and B. T. Ho
 1975 Role of Serotonin in the Discriminative Stimulus Properties of Mescaline. *Pharmacology, Biochemistry and Behavior* 3:429–35.

Bruhn, J. G.
 1971 Alcaloides en las cactaceas. *Cactaceas y Succulentas Mexicanas* 16: 51–58.
 1975 Pharmacognostic Studies of Peyote and Related Psychoactive Cacti. *Acta Universitatus Upsaliensis* 6:1–38.
 1977 Three Men and a Drug: Peyote Research in the 1890's. *Cactus and Succulent Journal of Great Britain* 39:27–30.

Bruhn, J. G., S. Agurell, and J. Lindgren
 1975 Cactaceae Alkaloids XXI. Phenethylamine Alkaloids of *Coryphantha* Species. *Acta Pharmaceutica Suecica* 12:199–204.

Bruhn, J. G., and C. Bruhn
 1973 Alkaloids and Ethnobotany of Mexican Peyote Cacti and Related Species. *Economic Botany* 27:241–51.

Bruhn, J. G., and B. Holmstedt
 1974 Early Peyote Research: An Interdisciplinary Study. *Economic Botany* 28:353–90.

Bruhn, J. G., and J. Lindgren
 1976 Cactaceae Alkaloids. XXIII. Alkaloids of *Pachycereus pecten-aboriginum* and *Cereus jamacaru*. *Lloydia* 39:175–77.

Bruhn, J. G., and J. Lundstrom
 1976 Alkaloids of *Carnegiea gigantea*. Arizonine, a New Tetrahydroisoquinoline Alkaloid. *Lloydia* 39:197–203.

Button Eaters.
 1959 *Time* 73 (16 February): 71.

Bye, R. A., Jr.
 1979a Hallucinogenic Plants of the Tarahumara. *Journal of Ethnopharmacology* 1:23–48.
 1979b An 1878 Ethnobotanical Collection From San Luis Potosí: Dr. Edward Palmer's First Major Mexican Collection. *Economic Botany* 33:135–62.
 1985 Medicinal Plants of the Tarahumara Indians of Chihuahua, Mexico. In *Two Mummies From Chihuahua, Mexico*, ed. R. A. Tyson and D. V. Elerick, 77–104. San Diego: San Diego Museum Papers No. 19, San Diego Museum of Man.

    1986   Medicinal Plants of the Sierra Madre: Comparative Study of the Tarahumara and Mexican Market Plants. *Economic Botany* 40: 103–24.

de Cárdenas, J.
    1945   *Primera parte de los problemas y secretos maravillosos de las Indias.* Ed. facsimilar de 1591. Madrid: Ediciones Cultura Hispanica.

Carlson, G. G., and V. H. Jones
    1939   Some Notes on Uses of Plants by the Comanche Indians. *Papers of the Michigan Academy of Science, Arts, and Letters* 25:517–42.

Cattabriga, A.
    1994   Propagation of Threatened Mexican Cacti. In *Threatened Cacti of Mexico*, by E. F. Anderson, S. Arias M., and N. P. Taylor, 117–33. Kew: Royal Botanic Gardens.

Cattell, J. P.
    1954   The Influence of Mescaline on Psychodynamic Material. *Journal of Nervous and Mental Disease* 119:233–44.

Cohen, S.
    1967   Psychotherapy with LSD: Pro and Con. In *The Use of LSD in Psychotherapy and Alcoholism*, ed. H. Alexander, 577–97. Indianapolis: Bobbs-Merrill.

Collier, D.
    1937   Peyote: A General Study of the Plant, the Cult, and the Drug. In *Survey of Conditions of the Indians in the United States.* Hearings before a Subcommittee of the Committee on Indian Affairs, U.S. Senate, 75th Cong., 1st sess., 18234–58. Washington, D.C.: U.S. Government Printing Office.

Correll, D. S., and M. C. Johnston
    1970   *Manual of the Vascular Plants of Texas.* Renner: Texas Research Foundation.

Coulter, J. N.
    1891   Manual of the Phanerogams and Pteridophytes of Western Texas. *Contributions from the U.S. National Herbarium* 2:1–588.
    1894   Preliminary Revision of the North American Species of *Cactus, Anhalonium,* and *Lophophora. Contributions from the U.S. National Herbarium* 3:91–132.

Cousineau, P.
    1993   Script for *The Peyote Road: Ancient Religion in Contemporary Crisis.* Produced and edited by Gary Rhine. 59 min. San Francisco: Kifaru Production. Videocassette.

Critchley, M.
  1972   Mescalism. *Aesculapius* 1:20–25.
Cronquist, A.
  1981   *An Integrated System of Classification of Flowering Plants.* New York: Columbia University Press.
Crosby, D. M., and J. L. McLaughlin
  1973   Cactus Alkaloids. XIX. Crystallization of Mescaline HCl and 3-methoxytyramine HCl from *Trichocereus pachanoi. Lloydia* 36: 416–18.
Davison, K.
  1976   Drug-Induced Psychoses and Their Relationship to Schizophrenia. In *Schizophrenia Today*, ed. D. Kemali, G. Bartholini, and D. Richter, 105–133. Oxford: Pergamon Press.
Dayish, Jr., F. J.
  1994   Testimony on 10 June before the Native American Affairs Subcommittee of the Natural Resources Committee, U.S. House of Representatives, Hearing on H.R. 4230, The American Indian Religious Freedom Act Amendments of 1994.
Denber, H.C.B., and S. Merlis
  1955   Studies on Mescaline VI. Therapeutic Aspects of the Mescaline-Chlorpromazine Combination. *Journal of Nervous and Mental Disease* 122:463–69.
Deniker, P.
  1957   Biological Changes in Man following Intravenous Administration of Mescaline. *Journal of Nervous and Mental Disease* 125:427–31.
Dishotsky, N. I., W. D. Loughman, R. E. Mogar, and W. R. Lipscomb
  1971   LSD and Genetic Damage. *Science* 172:431–40.
Ditman, K. S.
  1968   The value of LSD in psychotheraphy. In *The Problems and Prospects of LSD*, ed. J. Thomas, 45–60. Springfield, Ill.: Charles C. Thomas.
Dixon, W. E.
  1899/1900   The Physiological Action of the Alkaloids Derived from *Anhalonium Lewinii. Journal of Physiology* 25:69–86.
Dobkin de Rios, M.
  1968a   *Trichocereus pachanoi*: A Mescaline Cactus Used in Folk Healing in Northern Peru (San Pedro). *Economic Botany* 22:191–94.
  1968b   Folk Curing with a Psychedelic Cactus in Northern Peru. *Journal of Social Psychiatry* 15:23–32.

Dorrance, D. L., O. Janiger, and R. L. Teplitz
   1975   Effect of Peyote on Human Chromosomes. *Journal of the American Medical Association* 234:299–302.

Ellis, H.
   1897   A Note on the Phenomena of Mescal Intoxication. *Lancet* 1:1540–42.
   1898   Mescal: A New Artificial Paradise. *Contemporary Review* 73 (January): 130–41.
   1902   Mescal, a Study of a Divine Plant. *Popular Science Monthly* 61:52–71.

Expert Committee on Drugs Liable to Produce Addiction
   1950   *Report on the Second Session*. World Health Organization Technical Report Series No. 21. Geneva: World Health Organization.

Feigen, G. A., and G. A. Alles
   1955   Physiological Concomitants of Mescaline Intoxication. *Journal of Clinical and Experimental Psychopathology and Quarterly Review of Psychiatry and Neurology* 16:167–78.

Fernberger, S. W.
   1923   Observations on Taking Peyote (*Anhalonium lewinii*). *American Journal of Psychology* 34:267–70.

Fischer, R.
   1958   Pharmacology and Metabolism of Mescaline. *Review of Canadian Biology* 17:389–405.

Fisher, D. D.
   1968   The Chronic Side Effects from LSD. In *The Problems and Prospects of LSD*, ed. J. T. Ungerleider, 69–79. Springfield, Ill.: Charles C. Thomas.

Fisher, G.
   1965   Some Comments Concerning Dosage Levels of Psychedelic Compounds for Psychotherapeutic Experiences. In *The Psychedelic Reader*, ed. G. M. Weil, 149–59. New Hyde Park, N.Y.: University Books.

Frederking, W.
   1955   Intoxicant Drugs (Mescaline and Lysergic Acid Diethylamide) in Psychotherapy. *Journal of Nervous and Mental Disease* 121:262–66.

Fujimori, M., and H. S. Alpers
   1971   Psychotomimetic Compounds in Man and Animals. In *Biochemistry, Schizophrenics and Affective Illnesses*, ed. H. E. Himwich, 361–413 (Chap. 13). Baltimore: Williams and Wilkin.

Furst, P. T.
- 1971 *Ariocarpus retusus*, the 'False Peyote' of Huichol Tradition. *Economic Botany* 25:182–87.
- 1972a Introduction. In *Flesh of the Gods*, ed. P. T. Furst, vii–xvi. New York: Praeger.
- 1972b To Find Our Life: Peyote among the Huichol Indians of Mexico. In *Flesh of the Gods*, ed. P. T. Furst, 136–84. New York: Praeger.
- 1973 An Indian Journey to Life's Source. *Natural History* 82:34–43.
- 1974 Hallucinogens in Precolumbian Art. In *Art and Environment in Native America*, ed. M. E. King and I. R. Traylor, Jr., 55–101. The Museum Special Publication No. 7. Lubbock: Texas Tech University.
- 1975 Introduction. In *In the Magic Land of Peyote*, ed. F. Benitez, 11–23. New York: Warner Books.
- 1976 *Hallucinogens and Culture*. San Francisco: Chandler and Sharp.
- 1978 The Art of "Being Huichol." In *Art of the Huichol Indians*, ed. K. Berrin, 18–34. New York: Harry N. Abrams.
- 1989 Review of *Peyote Religion: A History*, by Omer C. Stewart. *American Ethnologist* 16: 386–87.
- 1995 "This Little Book of Herbs": Psychoactive plants as therapeutic agents in the Badianus manuscript of 1552. In *Ethnobatany: The Evolution of a Discipline*, ed. R. E. Schultes and S. von Reis, 108–30. Portland, Ore.: Dioscorides Press.

Furst, P. T., and M. D. Coe
- 1977 Ritual Enemas. *Natural History* 86:88–91.

Ganders, F. R.
- 1975 The Identity of *Echinocactus rapa*. *Cactus and Succulent Journal* (U.S.) 47:155–56.

García, Fr. B.
- 1760 *Manual para administrar los santos sacramentos etc.* Mexico City: Rivera.

Giarman, N. J., and D. X. Freedman
- 1965 Biochemical Aspects of the Actions of Psychotomimetic Drugs. *Pharmacological Reviews* 17:1–25.

Glennon, R. A., L. B. Kier, and A. T. Shulgin
- 1979 Molecular Connectivity Analysis of Hallucinogenic Mescaline Analogs. *Journal of Pharmaceutical Sciences* 68:906–7.

Goodman, L. S., and A. Gilman
- 1970 *The Pharmacological Basis of Therapeutics*. 4th ed. New York: Macmillan.

Greuter, W., H. M. Burdet, W. G. Chaloner, V. Demoulin, R. Grolle, D. L. Hawksworth, D. H. Nicolson, P. C. Silva, F. A. Stafleu, E. G. Voss, and J. McNeill, eds.
- 1988 *International Code of Botanical Nomenclature.* Regnum vegetabile vol. 118. Königstein: Koeltz Scientific Books.

Grinspoon, L.
- 1969 Marihuana. *Scientific American* 221:17–25.

Grollman, A.
- 1965 *Pharmacology and Therapeutics.* 6th ed. Philadelphia: Lea and Febiger.

Guttman, E.
- 1936 Artificial Psychoses Produced by Mescaline. *The Journal of Mental Science* 82:203–21.

Guttman, E., and W. S. Maclay
- 1936 Mescalin and Depersonalization. *Journal of Neurology and Psychopathology* 16:193–212.

Habermann, V.
- 1975 Two Red Flowering Species of *Lophophora. Cactus and Succulent Journal* (U.S.) 47:157–60.
- 1977 A Contribution to the Study of the Hallucinogenic Effect of Peyote (*Lophophora* Coulter). *Plzensky Lekarsky Sbornik* 44:17–22.
- 1978 Estimation of Mescaline and Pellotine in *Lophophora* Coulter Plants (Cactaceae) by Oscillographic Polarography. *Biokhimiya* 43:246–51.

Hardman, H. F., C. O. Haavik, and M. H. Seevers
- 1973 Relationship of the Structure of Mescaline and Seven Analogs to Toxicity and Behavior in Five Species of Laboratory Animals. *Toxicology and Applied Pharmacology* 25:299–309.

Harrisson, C.M.H., B. M. Page, and H. M. Keir
- 1976 Mescaline as a Mitotic Spindle Inhibitor. *Nature* 260:138–39.

Hayden, C.
- 1937 Speech of 4 August 1921, titled Indian Affairs–The Alleged Peyote Religion. In *Survey of Conditions of the Indians in the United States,* Hearings before a Subcommittee of The Committee on Indian Affairs, U.S. Senate, 75th Cong., 1st sess., 18278–18281. Washington, D.C.: U. S. Government Printing Office.

Heffter, A.
- 1894 Ueber Pellote. Ein Beitrage zur pharmakologischen Kenntniss der Cacteen. *Archiv für experimentelle Pathologie und Pharmakologie* 34: 65–86.

1896 Ueber Pellotin. *Therapeutische Monatshefte* 10:327–28.

1898 Ueber Pellote. Beiträge zur chemischen und pharmakologischen Kenntniss der Cacteen. Zweite Mittheilung. *Archiv für experimentelle Pathologie und Pharmakologie* 40:385–429.

Hennings, P.

1888 Eine giftige Kaktee, *Anhalonium lewinii* n. sp. *Gartenflora* 37:410–12.

Hernández, F.

1790 *De historia plantarum novae Hispaniae opera cum editatum inedita, ad autographi fidem et integratem expressa.* Matriti: Imp. Ibarra Heredum.

Hoebel, E. A.

1949 The Wonderful Herb, an Indian Cult Vision Experience. *Western Humanities Review* 3:126–30.

Hoffer, A., and H. Osmond

1967 *The Hallucinogens.* New York: Academic Press.

Hollister, L. E.

1962 Drug-induced Psychoses and Schizophrenic Reactions: A Critical Comparison. *Annals of the New York Academy of Sciences* 96:80–88.

1968 *Chemical Psychoses.* Springfield, Ill.: Charles C. Thomas.

1978 Psychotomimetic Drugs in Man. In *Handbook of Psychopharmacology*, ed. L. L. Iversen, S. D. Iversen, and S. N. Snyder, 389–424. New York: Plenum Press.

1982 Pharmacology and Toxicology of Psychotomimetics. In *Handbook of Experimental Pharmacology*, vol. 55/III, ed. F. Hoffmeister and G. Stille, 31–44. New York: Springer Verlag.

Hollister, L. E., and A, M. Hartman

1962 Mescaline, Lysergic Acid Diethylamide, and Psilocybin: Comparison of Clinical Syndromes, Effects on Color Perception, and Biochemical Measures. *Comprehensive Psychiatry* 3:235–41.

Holmgren, P. K., W. Keuken, and E. K. Schofield, comp.

1981 *Index Herbariorum.* Regnum Vegetabile, vol. 106. Antwerpt: Bohn, Scheltema and Holkema.

Hooker, W. J.

1847 Tab. 4296. *Echinocactus Williamsii. Curtis's Botanical Magazine* 73.

Hornemann, K.M.K., J. M. Neal, and J. L. McLaughlin

1972 Cactus Alkaloids XII: β-Phenethylamine Alkaloids of the Genus *Coryphantha. Journal of Pharmaceutical Sciences* 61:41–45.

Howard, J. H.

1957 The Mescal-bean Cult of the Central and Southern Plains: An

Ancestor of the Peyote Cult? *American Anthropologist* 59:75–87.

1962 Potawatomi Mescalism and Its Relationship to the Diffusion of the Peyote Cult. *Plains Anthropologist* 7:125–35.

Hurt, W. R.
1960 Factors in the Persistence of Peyote in the Northern Plains. *Plains Anthropologist* 5:16–27.

Huxley, A.
1956 Mescaline and the "Other World." In *Lysergic Acid Diethylamide and Mescaline in Experimental Psychiatry*, ed. L. Cholden, 46–50. New York: Grune and Stratton.

1959 *The Doors of Perception: Heaven and Hell*. Harmondsworth: Penguin Books.

Huxtable, R. J.
1992 The Pharmacology of Extinction. *Journal of Ethnopharmacology* 37: 1–11.

IOS Working Party
1990 The Genera of Cactaceae: Progress Towards Consensus. *Bradleya* 8: 85–107.

Jacobs, B. L.
1987 How Hallucinogenic Drugs Work. *American Scientist* 75:386–92.

Jacobsen, E.
1963 The Clinical Pharmacology of the Hallucinogens. *Clinical Pharmacology and Therapeutics* 4:480–503.

Jarvik, M. E.
1970a Drugs Used in the Treatment of Psychiatric Disorders. In *The Pharmacological Basis of Therapeutics*, ed. L. S. Goodman, and A. Gilman, 151–203. 4th ed. New York: Macmillan.

1970b Drugs, Hallucinations, and Memory. In *Origin and Mechanisms of Hallucinations*, ed. W. Keup, 277–301. New York: Plenum Press.

Joachimoglu, G., and E. Keeser
1924 Kakteen Alkaloide. In *Handbuch der Experimentellen Pharmakologie*, ed. A. Heffter, 1104–13. Berlin: Springer Verlag.

Joel, E.
1929 Peyotl. *Medizinische Welt* (Berlin) 3:265–67.

Joralemon, D.
1984 The Role of Hallucinogenic Drugs and Sensory Stimuli in Peruvian Ritual Healing. *Culture, Medicine, and Psychiatry* 8:399–430.

Kang, S., and J. P. Green
1970 Steric and Electronic Relationships among Some Hallucinogenic

Compounds. *Proceedings of the National Academy of Sciences* 67: 62–67.

Kapadia, G. J., and H. M. Fales
1968a Krebs Cycle Conjugates of Mescaline. Identification of Fourteen New Peyote Alkaloid Amides. *Chemical Communications* 1968:1688–89.
1968b Peyote Alkaloids VI. Peyophorine, a Tetrahydroisoquinoline Cactus Alkaloid Containing an N-ethyl Group. *Journal of Pharmaceutical Science* 57:2017–18.

Kapadia, G. J., and M.B.E. Fayez
1970 Peyote Constituents: Chemistry, Biogenesis, and Biological Effects. *Journal of Pharmaceutical Science* 59:1699–1727.
1973 The Chemistry of Peyote Alkaloids. *Lloydia* 36:9–35.

Kapadia, G. J., M.B.E. Fayez, B. K. Chowdhury, and H. M. Fales
1970 Structure and Synthesis of Peyoglunal, a New Cactus β-phenethylpyrrole. *Lloydia* 33:492.

Kapadia, G. J., and R. J. Highet
1968 Peyote Alkaloids IV. Structure of Peyonine, Novel β-phenethylpyrrole from *Lophophora williamsii*. *Journal of Pharmaceutical Science* 57: 191–92.

Kauder, E.
1899 Ueber Alkaloide aus *Anhalonium Lewinii*. *Archiv der Pharmazie* 237: 190–98.

Kelsey, F. E.
1959 The Pharmacology of Peyote. *South Dakota Journal of Medicine and Pharmacy* 12:231–33.

Kloesel, L.
1958 Some Notes on Peyote. *American Journal of Pharmacy* 130:307–16.

Klüver, H.
1966 *Mescal and Mechanisms of Hallucination*. Chicago: University of Chicago Press.

Knauer, A., and W.J.M.A. Maloney
1913 A Preliminary Note on the Psychic Action of Mescalin, with Special Reference to the Mechanism of Visual Hallucinations. *Journal of Nervous and Mental Disease* 40:397, 425–38.

Kolb, L. C.
1977 *Modern Clinical Psychiatry*. 9th ed. Philadelphia: W. B. Saunders.

La Barre, W.
1947 Primitive Psychotherapy in Native American Cultures: Peyotism and Confession. *Journal of Abnormal and Social Psychology* 42:294–309.

1957   Mescalism and Peyotism. *American Anthropologist* 59:708–11.
1960   Twenty Years of Peyote Studies. *Current Anthropology* 1:45–60.
1964   The 'Diabolic Root.' *The New York Times Magazine* (1 November): 96–98.
1972   Hallucinations and the Shamanic Origins of Religion. In *Flesh of the Gods*, ed. P. T. Furst, 261–78. New York: Praeger.
1979   Peyotl and Mescaline. *Journal of Psychedelic Drugs* 11:33–39.
1989   *The Peyote Cult*. 1938. 5th ed. New York: Schocken Books.

Landry, S. F.
1889   Notes on *Anhalonium Lewinii*, *Embelia ribes* and *Cocillana*. *Therapeutic Gazette* 3rd serial, 5:16.

Lanternari, V.
1965   *The Religions of the Oppressed*. New York: Mentor Books.

Lawson, P. E., and C. P. Morris
1991   The Native American Church and the New Court: The *Smith* Case and Indian Religious Freedom. *American Indian Culture and Research Journal* 15:79–91.

Laycock, D.
1990   Watering Down the Free-Exercise Clause. *The Christian Century* 107: 518–19.

Lemaire, C.
1839   *Cactearum Genera Nova et Species Nova en Horto Monville*. Paris: Lutetiae Parisiorum.

Lemberger, L., and A. Rubin
1976   *Physiological Disposition of Drugs of Abuse*. New York: Spectrum Publications.

Lennard, H. L., L. J. Epstein, A. Bernstein, and D. C. Ransom
1970   Hazards Implicit in Prescribing Psychoactive Drugs. *Science* 169: 438–41.

Leonard, I.
1942   Peyote and the Mexican Inquisition. *American Anthropologist* 44: 324–26.

Leuenberger, B. E.
1976   *Die Pollenmorphologie der Cactaceae*. Dissertationes Botanicae, No. 31. Vaduz: J. Cramer.

Leuner, H.
1963   Psychotherapy with Hallucinogens. In *Hallucinogenic Drugs and Their Psychotherapeutic Use*, ed. R. Crocket, R. A. Sandison, and A. Walk, 67–73. Springfield, Ill.: Charles C. Thomas.

Lewin, L.
    1888a Anhalonium Lewinii. *Therapeutic Gazette* 3rd serial, 4:231–37.
    1888b Ueber *Anhalonium Lewinii*. *Archiv für experimentelle Pathologie und Pharmakologie* 24:401–11.
    1964 *Phantastica: Narcotic and Stimulating Drugs*. Trans. of 1927 German edition. New York: E. P. Dutton.

Li, H. L., and J. J. Willaman
    1968 Distribution of Alkaloids in Angiosperm Phylogeny. *Economic Botany* 22:239–52.

Loftin, J. D.
    1989 Anglo-American Jurisprudence and the Native American Tribal Quest for Religious Freedom. *American Indian Culture and Research Journal* 13:1–52.

Ludwig, A. M.
    1969 Altered States of Consciousness. In *Altered States of Consciousness*, ed. C. C. Tart, 9–22. New York: John Wiley and Sons.

Ludwig, A. M., and J. Levine
    1966 The Clinical Effects of Psychedelic Agents. *Clinical Medicine* 73:21–24.

Lumholtz, C.
    1900 Symbolism of the Huichol Indians. *Memoirs of the American Museum of Natural History* 3:1–228.
    1902 *Unknown Mexico*. 2 vols. New York: Charles Scribner's Sons.

Ma, W. W., X. Y. Jiang, R. G. Cooks, J. L. McLaughlin, A. C. Gibson, F. Zeylemaker, and C. N. Ostolaza
    1986 Cactus Alkaloids, LXI. Identification of Mescaline and Related Compounds in Eight Additional Species Using TLC and MS/MS. *Journal of Natural Products* 49:735–37.

Malitz, S., B. Wilkens, and H. Esecover
    1962 A Comparison of Drug-induced Hallucinations with Those Seen in Spontaneously Occurring Psychoses. In *Hallucinations*, ed. L. J. West, 50–63. New York: Grune and Stratton.

Malouf, C.
    1942 Gosiute Peyotism. *American Anthropologist* 44:93–103.

Marks, L. E.
    1975 On Colored-Hearing Synesthesia: Cross-Modal Translations of Sensory Dimensions. *Psychological Bulletin* 82:303–31.

Marriott, A., and C. K. Rachlin
    1971 *Peyote*. New York: Thomas Y. Crowell.

Mason, C. T., Jr., and P. B. Mason
  1987 A Handbook of Mexican Roadside Flora. Tucson: University of Arizona Press.
Mata, R., J. L. McLaughlin, and W. H. Earle
  1976 Cactus Alkaloids: XXX. N-methylated tyramines from *Trichocereus spachianus, T. candicano,* and *Espostoa huanucensis. Lloydia* 39: 461–63.
McAllester, D. P.
  1949 *Peyote Music.* New York: Viking Fund Publications in Anthropology, No. 13.
McCall, R. B.
  1986 Effects of Hallucinogenic Drugs on Serotonergic Neuronal Systems. *Pharmacology Biochemistry and Behavior* 24:359–63.
McCleary, J. A., P. S. Sypherd, and D. L. Walkington
  1960 Antibiotic Activity of an Extract of Peyote (*Lophophora williamsii* (Lemaire) Coulter). *Economic Botany* 14:247–49.
McCleary, J. A., and D. L. Walkington
  1964 Antimicrobial Activity of the Cactaceae. *Bulletin of the Torrey Botanical Club* 91:361–69.
McKern, S. S., and T. W. McKern
  1970 The Peace Messiah. *Mankind* 2:59–69.
McLain, L.
  1968 A Study of Peyote. *Clinical Toxicology* 1:81–85.
McLaughlin, J. L.
  1973 Peyote: An Introduction. *Lloydia* 36:1–8.
McLaughlin, J. L., and A. G. Paul
  1965 Presence of Hordenine in *Lophophora williamsii. Journal of Pharmaceutical Sciences* 54:661.
  1966 The Cactus Alkaloids. I. Identification of N-methylated Tyramine Derivatives in *Lophophora williamsii. Lloydia* 29:315–27.
  1967 The Cactus Alkaloids. II. Biosynthesis of Hordenine and Mescaline in *Lophophora williamsii. Lloydia* 30:91–99.
Merrill, W. L.
  1977 *An Investigation of Ethnographic and Archaeological Specimens of Mescalbeans (Sophora Secundiflora) in American Museums.* Technical Report No. 6, Research Reports in Ethnobotany, contribution 1, Ann Arbor: Museum of Anthropology, University of Michigan.
Miller, R.
  1988 Homegrown Peyote. *High Times* 1988:36–39.

Mitchell, S. W.
    1896   The Effects of Anhalonium Lewinii (the Mescal Button). *British Medical Journal* 2:1625–28.

de Molina, A.
    1880   *Vocabulario de la lengua Mexicana.* Facsimile of 1571 edition. Leipzig: B. G. Teubner.

Mooney, J.
    1896   The Mescal Plant and Ceremony. *Therapeutic Gazette*, 3rd serial 20: 7–11.

Morgan, G. R.
    1983a   Hispano-Indian Trade of an Indian Ceremonial Plant, Peyote (*Lophophora williamsii*), on the Mustang Plains of Texas. *Journal of Ethnopharmacology* 9:319–21.
    1983b   The Biogeography of Peyote in South Texas. *Botanical Museum Leaflets, Harvard University* 29:73–86.

Morgan, G., and O. C. Stewart
    1984   Peyote Trade in South Texas. *Southwestern Historical Quarterly* 87: 269–96.

Mothes, K., and M. Luckner
    1985   Historical Introduction. In *Biochemistry of Alkaloids*, ed. K. Mothes, H. R. Schütte, and M. Luckner, 15–20. Weinheim: VCH Publishers.

Mount, G., compiler and editor
    1987   *The Peyote Book.* Arcata, Calif.: Sweetlight Books.

Muller, C. H.
    1947   Vegetation and Climate of Coahuila, Mexico. *Madroño* 9:33–57.

Muller, K.
    1978   Huichol Art and Acculturation. In *Art of the Huichol Indians*, ed. K. Berrin, 84–100. New York: Harry N. Abrams.

Murie, J. R.
    1914   Pawnee Indian Societies. *Anthropological Papers, American Museum of Natural History* 11 (pt. 7): 543–644.

Myerhoff, B. G.
    1974   *Peyote Hunt: The Sacred Journey of the Huichol Indians.* Ithaca, N.Y.: Cornell University Press.
    1978   Peyote and the Mystic Vision. In *Art of the Huichol Indians*, ed. K. Berrin, 56–70. New York: Harry N. Abrams.

Naditch, M. P.
    1974   Acute Adverse Reactions to Psychoactive Drugs, Drug Usage, and Psychopathology. *Journal of Abnormal Psychology* 83:394–403.

National Clearinghouse for Drug Abuse Information
- 1973 *Mescaline.* U.S. Department of Health, Education, and Welfare Publication No. (ADM) 75–204. (Formerly NCDAI Publication No. 19.)

Neal, J. M., P. T. Sato, W. N. Howald, and J. L. McLaughlin
- 1972 Peyote Alkaloids: Identification in the Mexican Cactus *Pelecyphora aselliformis* Ehrenberg. *Science* 176:1131–33.

Neal, J. M., P. T. Sato, and J. L. McLaughlin
- 1971 Cactus Alkaloids. XI. Isolation of Tyramine, N-Methyltyramine, and Hordenine from *Obregonia denegrii. Economic Botany* 25:382–84.

Nettl, B.
- 1953 Observations on Meaningless Peyote Song Texts. *Journal of American Folklore* 66:161–64.

Newberne, R.E.L.
- 1925 *Peyote, an Abridged Compilation from the Files of the Bureau of Indian Affairs.* 3d ed., revised and corrected. Lawrence, Kans.: Haskell Printing Department, Haskell Institute.

Nichols, D. E.
- 1981 Structure-Activity Relationships of Phenethylamine Hallucinogens. *Journal of Pharmaceutical Sciences* 70:839–49.

Nichols, D. E., W. R. Pfister, G.K.W. Yim, and R. J. Cosgrove
- 1977 A New View of the Structural Relationship Between LSD and Mescaline. *Brain Research Bulletin* 2:169–71.

Ochoterena, I.
- 1922 *Las cactáceas de México.* Mexico City: Editorial "Cultura."

Opler, M. E.
- 1936 The Influence of Aboriginal Pattern and White Contact on a Recently Introduced Ceremony, the Mescalero Peyote Cult. *Journal of American Folklore* 49:143–66.
- 1938 The Use of Peyote by the Carrizo and Lipan Apache Tribes. *American Anthropologist* 40:271–85.
- 1939 A Description of a Tonkawa Peyote Meeting Held in 1902. *American Anthropologist* 41:433–39.

Opler, M. K.
- 1940 The Character and History of the Southern Ute Peyote Rite. *American Anthropologist* 42:463–78.

Ortega, J.
- 1887 *Historia del Nayarit, Sonora, Sinaloa, y ambas Californias.* 1754. Mexico City: E. Abadiano.

Osmond, H.
- 1970 On Being Mad. In *Psychedelics*, ed. B. Aaronson and H. Osmond, 21–28. Garden City, N.Y.: Doubleday.

Osmond, H., and J. Smythies
- 1952 Schizophrenia–A New Approach. *Journal of Mental Science* 98: 309–15.

Ott, J.
- 1976 *Hallucinogenic Plants of North America*. Berkeley, Calif.: Wingbow.

Padula, P. A., and L. W. Friedmann
- 1987 Acquired Amputation and Prostheses Before the Sixteenth Century. *Angiology* 38:133–41.

Pascarosa, P., S. Futterman, and M. Halsweig
- 1976 Observations of Alcoholics in the Peyote Ritual: A Pilot Study. *Annals of the New York Academy of Sciences* 273:518–24.

de Pasquale, A.
- 1984 Pharmacognosy: The Oldest Modern Science. *Journal of Ethnopharmacology* 11:1–16.

Pelletier, S. W.
- 1983 The Nature and Definition of an Alkaloid. In *Alkaloids: Chemical and Biological Perspectives*, vol. 1, ed. S. W. Pelletier. New York: John Wiley and Sons.

Petrullo, V.
- 1934 *The Diabolic Root: A Study of Peyotism the New Indian Religion, among the Delawares*. Philadelphia: University of Pennsylvania Press.

Pfeiffer, C. C., and H. B. Murphree
- 1965 Introduction to Psychotropic Drugs and Hallucinogenic Drugs. In *Drill's Pharmacology in Medicine*, ed. J. R. DiPalma, 321–36. 3d ed. New York: McGraw-Hill.

Pierson, D. L.
- 1915 American Indian Peyote Worship. *Missionary Review* 38:201–6.

Prentiss, D. W., and F. P. Morgan
- 1895 *Anhalonium lewinii* (Mescal Buttons). *Therapeutic Gazette* 3d serial, 11:577–85.
- 1896 Therapeutic Uses of Mescal Buttons (*Anhalonium Lewinii*). *Therapeutic Gazette* 3d serial, 12:4–7.

Prescott, W. H.
- 1855 *History of the Conquest of Mexico*. 3 vols. Boston: Phillips, Sampson and Company.

Prieto, A.
- 1873 *Historia, geografía, y estadistica del estado de Tamaulipas.* Mexico City: Escalerillas.

Radin, P.
- 1914 A Sketch of the Peyote Cult of the Winnebago: A Study in Borrowing. *American Journal of Religious Psychology and Education* 7:1–22.
- 1923 The Winnebago Tribe. *Annual Report, Bureau of American Ethnology* 37:35–550.
- 1926 *Crashing Thunder: The Autobiography of a Winnebago Indian.* New York: Appleton.
- 1963 *The Autobiography of a Winnebago Indian.* New York: Dover.

Ramirez, F. A., and A. Burger
- 1950 The Reduction of Phenolic β-nitro-styrenes by Lithium Aluminum Hydride. *Journal of the American Chemical Society* 72:2781–82.

Ranieri, R. L., and J. L. McLaughlin
- 1975 Cactus Alkaloids XXVII. Use of Fluorescamine as a Thin-layer Chromatographic Visualization Reagent for Alkaloids. *Journal of Chromatography* 111:234–37.

Rao, G. S.
- 1970 Identity of Peyocactin, an Antibiotic from Peyote (*Lophophora williamsii*) and Hordenine. *Journal of Pharmacy and Pharmacology* 22:544–45.

Rave, H.
- 1937 Statement read into Congressional Record on 11 October 1911. In *Survey of Conditions of the Indians in the United States.* Hearings before a Subcommittee of the Committee on Indian Affairs, U.S. Senate, 75th Cong., 1st sess., 18264–66. Washington, D.C.: U.S. Government Printing Office.

Rech, R. H., and R. L. Commissaris
- 1982 Neurotransmitter Basis of the Behavioral Effects of Hallucinogens. *Neuroscience & Biobehavioral Reviews* 6:521–27.

Reti, L.
- 1950 Cactus Alkaloids and Some Related Compounds. *Fortschritte der Chemie Organischer Naturstoffe* 6:242–89.
- 1953 Beta-phenethylamines. In *The Alkaloids*, ed. R.H.F. Manske and H. L. Holmes, 3:313–38. New York: Academic Press.
- 1954 Simple Isoquinoline Alkaloids. In *The Alkaloids*, ed. R.H.F. Manske, and H. L. Holmes, 4:7–21. New York: Academic Press.

Reynolds, P. C., and E. J. Jindrich
- 1985 A Mescaline Associated Fatality. *Journal of Analytical Toxicology* 9: 183–84.

Rhodes, W.
- 1958 A Study of Musical Diffusion Based on the Wandering of the Opening Peyote Song. *International Folk Music Journal* 10:42–49.

Richardson, D. A.
- 1896 A Report on the Action of *Anhalonium Lewinii* (Mescal Button). *New York Medical Journal* 64:194–95.

Robinson, T.
- 1967 *The Organic Constituents of Higher Plants*. Minneapolis: Burgess.
- 1974 Metabolism and Function of Alkaloids in Plants. *Science* 184:430–35.
- 1981 *The Biochemistry of Alkaloids*. 2d ed. Berlin: Springer Verlag.

Roseman, B.
- 1963 *The Peyote Story*. Hollywood, Calif.: Wilshire Book Co.

Rouhier, A.
- 1927 *La Plante qui fait les yeux émerveillés: Le Peyotl*. Paris: Gaston Doin et Cie.

Rümpler, T.
- 1886 *Carl Friedrich Förster's Handbuch der Cacteenkunde*. Leipzig: Im. Tr. Woller.

Rzedowski, J.
- 1966 Vegetación del estado San Luis Potosí. *Acta Científica Potosina* 5: 1–291.
- 1978 *Vegetación de México*. Mexico City: Limusa.

Safford, W. E.
- 1915 An Aztec Narcotic (*Lophophora williamsii*). *Journal of Heredity* 6: 291–311.

de Sahagún, Fr. B.
- 1938 *Historia general de las cosas de Nueva España*. Vol. 3. Mexico City: Pedro Robredo.

Salm-Dyck, J. von
- 1845 Beschreibung einiger neuen Cacteen. *Allgemein Gartenzeitung* 13: 385–86.

Schaefer, S. B.
- 1996 The Crossing of the Souls: Peyote, Perception and Meaning Among the Huichol Indians. In *People of the Peyote*, ed. S. B. Schaefer and P. T. Furst. Albuquerque: University of New Mexico Press, in press.

Scheibel, M. E., and A. B. Scheibel
- 1962 Hallucinations and the Brain Stem Reticular Core. In *Hallucinations*, ed. L. J. West, 15–35. New York: Grune and Stratton.

Schendel, G.
- 1968 *Medicine in Mexico*. Austin: University of Texas Press.

Schlee, D.
- 1985 The Ecological Significance of Alkaloids. In *Biochemistry of Alkaloids*, ed. K. Mothes, H. R. Schütte, and M. Luckner, 56–64. Deerfield Beach, Fla.: VCH Publishers.

Schultes, R. E.
- 1937 Peyote (*Lophophora williamsii*) and Plants Confused with It. *Harvard University Botanical Museum Leaflets* 5:61–88.
- 1938 The Appeal of Peyote (*Lophophora williamsii*) as a Medicine. *American Anthropologist* 40:698–715.
- 1940 The Aboriginal Therapeutic Uses of *Lophophora williamsii*. *Cactus and Succulent Journal* (U.S.) 12:177–81.
- 1969 Hallucinogens of Plant Origin. *Science* 163:245–54.
- 1970 The Botanical and Chemical Distribution of Hallucinogens. *Annual Review of Plant Physiology* 21:571–98.

Schultes, R. E., and A. Hofmann
- 1979 *Plants of the Gods*. New York: McGraw-Hill.
- 1980 *The Botany and Chemistry of Hallucinogens*. 2d ed. Springfield, Ill.: Charles C. Thomas.

Schultes, R. E., W. M. Klein, T. Plowman, and T. E. Lockwood
- 1974 *Cannabis*: An Example of Taxonomic Neglect. *Harvard University Botanical Museum Leaflets* 23:337–67.

Schütte, H. R., and H. W. Liebisch
- 1985 Alkaloids Derived from Tyrosine and Phenylalanine. In *Biochemistry of Alkaloids*, ed. K. Mothes, H. R. Schütte, and M. Luckner, 106–27. Deerfield Beach, Fla.: VCH Publishers.

Schwartz, Richard H.
- 1988 Mescaline: A Survey. *American Family Physician* 37:122–24.

Seevers, M. H.
- 1958 Drug Addiction. In *Pharmacology in Medicine*, ed. V. A. Drill, 236–52. 2d ed. New York: McGraw-Hill.

Sharon, D.
- 1972 The San Pedro Cactus in Peruvian Folk Healing. In *Flesh of the Gods*, ed. P. T. Furst, 114–35. New York: Praeger Press.

Shemluck, M.
- 1982 Medicinal and Other Uses of the Compositae by Indians in the United States and Canada. *Journal of Ethnopharmacology* 5:303–58.

Shonle, R.
- 1925 Peyote: The Giver of Visions. *American Anthropologist* 27:53–75.

Shulgin, A. T.
- 1970 Chemistry and Structure-Activity Relationships of the Psychotomimetics. In *Psychotomimetic Drugs*, ed. D. H. Efron, 21–41. New York: Raven Press.
- 1973 Mescaline: The Chemistry and Pharmacology of Its Analogs. *Lloydia* 36:46–58.
- 1978 Psychotomimetic Drugs: Structure-Activity Relationships. In *Handbook of Psychopharmacology*, ed. L. L. Iversen, S. D. Inversen, and S. H. Snyder, 243–333. New York: Plenum Press.
- 1979a Chemistry of Phenethylamines Related to Mescaline. *Journal of Psychedelic Drugs* 11:41–52.
- 1979b Profiles of Psychedelic Drugs. 7. Mescaline. *Journal of Psychedelic Drugs* 11:355.

Siegel, R. K., P. R. Collings, and J. L. Díaz
- 1977 On the Use of *Tagetes lucida* and *Nicotiana rustica* as a Huichol Smoking Mixture: the Aztec "Yahutli" with Suggestive Hallucinogenic Effects. *Economic Botany* 31:16–23.

Simpson, L., and P. McKellar
- 1955 Types of Synaesthesia (Mescal Experiments). *Journal of Mental Science* 101:141–47.

Sinnett, E. R.
- 1970 Experience and Reflections. In *Psychedelics*, ed. B. Aaronson and H. Osmond, 29–35. Garden City, N.Y.: Doubleday.

Skinner, A.
- 1915 Societies of the Iowa, Kansa, and Ponca Indians. *Anthropological Papers, American Museum of Natural History* 11:679–740.

Slotkin, J. S.
- 1952 Menomini Peyotism. *Transactions of the American Philosophical Society* 42:565–700.
- 1955 Peyotism, 1521–1891. *American Anthropologist* 57:202–30.
- 1956 *The Peyote Religion*. Glencoe, Ill.: Free Press.

de Smet, P.A.G.M.
- 1983 A Multidisciplinary Overview of Intoxicating Enema Rituals in the Western Hemisphere. *Journal of Ethnopharmacology* 9:129–66.

de Smet, P.A.G.M., and F. J. Lipp, Jr.
- 1987 Supplementary Data on Ritual Enemas and Snuffs in the Western Hemisphere. *Journal of Ethnopharmacology* 19:327–31.

Smith, A.
- 1993 Interview in *The Peyote Road: Ancient Religion in Contemporary Crisis*. Produced and edited by Gary Rhine. 59 min. San Francisco: Kifaru Production. Videocassette.

Smythies, J. R.
- 1963 The Mode of Action of Mescaline. In *Hallucinogenic Drugs and Their Psychotherapeutic Use*, ed. R. Crocket, R. A. Sandison, and A. Walk, 17–22. Springfield, Ill.: Charles C. Thomas.

Smythies, J. R., R. J. Bradley, V. S. Johnston, F. Benington, R. D. Morin, and L. C. Clark, Jr.
- 1967 Structure-activity Relationship Studies on Mescaline: III. The Influence of the Methoxy Groups [Rat]. *Psychopharmacologia* 10:379–87.

Smythies, J. R., E. A. Sykes, and C. P. Lord
- 1966 Structure-activity Relationship Studies on Mescaline: II. Tolerance and Cross-tolerance between Mescaline and Its Analogues in the Rat. *Psychopharmacologia* 9:434–46.

Snake, R.
- 1993 Interview in *The Peyote Road: Ancient Religion in Contemporary Crisis*. Produced and edited by Gary Rhine. 59 min. San Francisco: Kifaru Production. Videocassette.

Soto Mora, C., and E. Jáuregui O.
- 1965 *Isotermas extremas e indice de aridez en la República Mexicana*. Mexico City: Universidad Nacional Autónoma de México.

Späth, E.
- 1919 Ueber die *Anhalonium*-Alkaloide. *Monatschefte für Chemie und verwandte Teile anderer Wissenschaften* 40:129–54.

Speck, F. G.
- 1933 Notes on the Life of John Wilson, the Revealer of Peyote, as Recalled by His Nephew, George Anderson. *General Magazine and Historical Chronicle* 35:539–56.

Speck, L. B.
- 1957 Toxicity and Effects of Increasing Doses of Mescaline. *Journal of Pharmacology and Experimental Therapeutics* 119:78–84.

Spindler, G. D.
- 1952 Personality and Peyotism in Menomini Indian Acculturation. *Psychiatry* 15:151–59.

Stenberg, M. P.
- 1946 The Peyote Cult Among Wyoming Indians. *University of Wyoming Publications* 12:85–156.

Stewart, O. C.
- 1944 Washo-northern Paiute Peyotism: A Study in Acculturation. *University of California Publications in American Archeology and Ethnology* 40:63–141.
- 1956 Peyote and Colorado's Inquisition Law. *Colorado Quarterly* 5:79–90.
- 1974 Origin of the Peyote Religion in the United States. *Plains Anthropologist* 19:211–23.
- 1980 Peyotism and Mescalism. *Plains Anthropologist* 25:297–309.
- 1984 Friend to the Ute. In *Peyotism in the West*, ed. O. C. Stewart and D. F. Aberle, 269–75. University of Utah Anthropological Papers, No. 108. Salt Lake City: University of Utah Press.
- 1987 *Peyote Religion: A History*. Norman: University of Oklahoma Press.

Superweed, M. J.
- 1968 *The Marijuana Consumer's and Dealer's Guide with the New Superior Mescaline Process*. San Francisco: Stone Kingdom Syndicate.

Swan, G. A.
- 1967 *An Introduction to the Alkaloids*. New York: John Wiley and Sons.

Szara, S.
- 1972 The Hallucinogenic Drugs–Curse or Blessing? In *Drug Abuse: Medical and Criminal Aspects*, ed. F. Braceland, et al., 215–20. New York: MSS Information Corporation.

Taylor, N.
- 1949 *Flight from Reality*. New York: Duell, Sloan, and Pearce.

Teitelbaum, D., and D. Wingeleth
- 1977 Diagnosis and Management of Recreational Mescaline Self Poisoning. *Journal of Analytical Toxicology* 1:36–37.

Tepker, H. F., Jr.
- 1991 Hallucinations of Neutrality in the Oregon Peyote Case. *American Indian Law Review* 16:1–56.

Thomas, E.
- 1937 Statements in Congressional Record. In *Survey of Conditions of the Indians in the United States*. Hearings before a Subcommittee of the

  Committee on Indian Affairs, U.S. Senate, 75th Cong., 1st sess., 18167, 18231. Washington, D.C.: U.S. Government Printing Office.

Todd, J. S.
 1969 Thin-layer Chromatography Analysis of Mexican Populations of *Lophophora* (Cactaceae). *Lloydia* 32:395–98.

Troike, R.
 1962 The Origin of Plains Mescalism. *American Anthropologist* 64:946–63.

Tsao, M.
 1951 A New Synthesis of Mescaline. *Journal of the American Chemical Society* 73:5495–96.

Underhill, R.
 1952 Peyote. In *Proceedings of the 30th International Congress of Americanists*, 143–48. London.
 1957 Religion among American Indians. *Annals of the American Academy of Sciences* 311:127–36.

Unger, S. M.
 1963 Mescaline, LSD, Psilocybin, and Personality Change: A Review. *Psychiatry* 26:111–25.

U.S. Congress
 1970 Public Law 91–513. 91st Congress, H. R. 18583. October 27, 1970. 84 STAT. 1236.

Vite-González, F., J. A. Zavala-Hurtado, M. A. Armella-Villalpando, and M. D. García-Suárez
 1990 Regionalización y caracterización macroclimática del matorral xerófilo. Tópicos Fitogeográficos, IV.8.3. *Atlas Nacional de México*. Mexico City: Instituto de Geografía, Universidad Nacional Autónoma de México.

Voss, A.
 1894 Genus 427. *Ariocarpus* Scheidw. Aloecactus. In *Vilmorin's Blumengartnerei*, ed. A. Siebert, 368. Berlin.

Wagner, R. W.
 1975 Pattern and Process in Ritual Syncretism: The Case of Peyotism Among the Navajo. *Journal of Anthropological Research* 31:162–81.

Waller, G. R., and E. K. Nowacki
 1978 *Alkaloid Biology and Metabolism in Plants*. New York: Plenum Press.

Wasson, R. G.
 1965 Notes on the Present Status of Ololiuhqui and the Other Hallucinogens of Mexico. In *The Psychedelic Reader*, ed. G. M. Weil, 163–89. New Hyde Park, N.Y.: University Books.

Wells, B.
  1973 *Psychedelic Drugs.* Baltimore: Penguin Books.
West, L. G., R. L. Vanderveen, and J. L. McLaughlin
  1974 β-Phenethylamines from the Genus *Gymnocactus. Phytochemical Reports* 13:665–66.
Wikler, A.
  1957 *The Relation of Psychiatry to Pharmacology.* Baltimore: Williams and Wilkins.
Willaman, J. J., and B. G. Schubert
  1961 *Alkaloid-bearing Plants and Their Contained Alkaloids.* U.S. Department of Agriculture Technical Bulletin No. 1234. Washington, D.C.: U.S. Government Printing Office.
Wyatt, R. J., E. H. Cannon, D. M. Stoff, and J. C. Gillin
  1976 Interactions of Hallucinogens at the Clinical Level. *Annals New York Academy of Sciences* 281:456–86.
Zaehner, R. C.
  1961 *Mysticism, Sacred and Profane.* London: Oxford University Press.
Zimmerman, A. D.
  1985 Systematics of the Genus *Coryphantha* (Cactaceae). Ph.D. diss., University of Texas at Austin.

Index

Aberle, David F., 66
Acacia, 173; *A. farnesiana*, 174; *A. greggii*, 174; *A. rigidula*, 174
acetylcholine, 123
N-acetyl-3,4-dimethoxy-5-hydroxyphenylethylamine, 124
N-acetylmescaline, 124, 138, 224
acute brain disorders. *See* delirium
addicting liability, 186
addiction, 131, 185–87
Agaricaceae, 129, 164–65
Agave, 206; *A. angustifolia*, 163; cactus, 173; *A. lechuguilla*, 16, 169, 172, 173–75; *A. striata*, 173; *A. stricta*, 173
Agurell, Stig, 150
Aiken, John and Luisa, 197
Alarcón, Hernando Ruíz de, 20
alcohol and alcoholism, 110–11, 184
Alegre, Francisco Javier, 21
Alex, Tom, 169
alkaloidal amines, 141
alkaloids, 83, 107, 115–16, 118, 120–22, 132, 162, 206; and angiosperm phylogeny, 136; biosynthesis of, 133; defined, 133–34; functions of, 135. *See also* peyote, alkaloids of
Alles, Gordon A., xiii, xiv, 154
"altered state of consciousness," 80
American Indian Relgious Freedom Act, 201, 204–5, 236–37; amendments to, 205, 238–39
amino acids, 141
amphetamine, 149–50
anacahuite, 174
anesthetics, 5
angel's trumpet, 5
anhalamine, 138, 141, 144, 152, 224
anhalidine, 121, 138, 141–42, 152, 225
anhalinine, 133–34, 138, 140–41, 144, 225
anhalonidine, 121, 138, 141, 143–44, 226

anhalonine, 121, 141, 143–44, 152, 154, 227
Anhalonium, 111–14, 138, 153, 155; *A. jourdanianum*, 216; *A. lewinii*, 114, 157, 213–14; *A. rungei*, 218; *A. subnodosum*, 218; *A. visnagra*, 219; *A. williamsii*, 156–57, 212
Areca catechu, 184
Ariocarpus, 142, 153, 155, 157, 166–67, 178; *A. agavoides*, 142–43, 162; *A. fissuratus*, 162; *A. kotschoubeyanus*, 162; *A. retusus*, 18, 162, 173; *A. williamsii*, 212
Arizona Constitution, 195
Arlegui, Joseph de, 13
Artemisia spp., 55
Association of Mescal Bean Eaters, 47
Asteraceae, 160, 163
Astrophytum; *A. asterias*, 162; *A. capricorne*, 162; *A. myriostigma*, 162, 173
atlinan, 20
atropine, 135
Attakai, Mary, 195
Aztecs, 4–6, 160, 164
Aztekium, 168, 178; *A. ritterii*, 162

Badiano, Juan, 5
Badianus manuscript, 5
barbiturates, 126, 187
barley, 121
Bates, Marston, 184, 187
belladonna, 112, 123
Bennett, Wendell Clark, 22, 109
Benson, Lyman, xiii–xiv
Bergman, Robert L., 193
Beringer, Kurt, 81
Bernstein, Arnold, 184
betel nut, 184
"Big Moon" ceremony, 46, 64
biote, 109
Bishop's cap cactus, 173

263

black brush, 174
Blofeld, John, 104
"borrados," 23
Bosque Espinoso, 174
Braden, William, 85, 88, 98
Bronze Age, 3
Briggs, John R., 155, 214
Bruhn, Jan G., 136, 144, 213, 218
Bureau of Indian Affairs, 42, 125, 163, 194
Bursera fagaroides, 174
Bye, Robert, 21–22

Cacalia cordifolia, 160, 163
Cactaceae, 154, 177, 209
Cacteae, 167
Cactoideae, 167, 177–78
caffeine, 113, 135
California Supreme Court, 196
camphor monobromide, 113
candicine, 140, 220
Cannabaceae, 131, 165
Cannabis, 131, 165; *C. indica*, 112–13; *C. sativa*, 198
Cárdenas, Amada, 196
Cárdenas, Juan de, 7–9
cardiac glycosides, 123
Carnegiea gigantea, 143
Caryophyllales, 177
cat claw, 173–74
Celtis pallida, 174
cenizo, 174
Cercidium macrum, 174
ceremony of the sacred stones, 31, 34
Cereus, 142
Chavin culture of Peru, 21
Chihuahuan Desert, 33, 168, 170–75, 206; Chihuahuan Desert Shrub, 172
chloral, 113
chlorpromazine, 118
cholla, 174
Christianity: attitudes toward peyote, 34–36, 63, 183, 188–93; denominations influencing Indians, 36; influence on peyotism, 28, 32, 39, 59–60, 62–64; inquisition, 9, 188; missionaries, 35, 42, 60, 63, 163, 189–93; Protestantism and Indians, 47, 65, 189–92; Roman Catholicism and Indians, 5, 11, 23, 33, 67, 188; symbolism, 25, 30
Church of Jesus Christ of Latter-Day Saints (Mormon), 47
Church of the Awakening, 197
citrimide, 141, 229
Claremont Graduate School, xiii
Clark, David S., 196
clepe, 175
coca, 200, 233
cocaine, 127, 135, 184, 187
codeine, 134
Coe, M. D., 21
colcemid, 129
colchicine, 129, 134
Colima culture, 3
Collier, Donald, 193
Collier, John, 194
Compositae. *See* Asteraceae
Comprehensive Drug Abuse Prevention and Control Act of 1970, 197–201, 205, 232–36
Condalia lycioides, 173
Conocybe, 164–65
Convolvulaceae, 160, 164–65
Cordia boissieri, 174
Cortéz, Hernán, 4
Coryphantha, 142, 173
Cotyledon caespitosa, 160
Coulter, John, 157
Cowper, Dennis, 215
coyotillo, 174
Crassulaceae, 160, 163
creosote bush, 17, 169, 171, 173, 175
cross-tolerance, 128
crucifixion thorn, 173
Cruz, Martín de la, 5
curandero, 4, 11, 107
cytisine, 32, 163

Dasylirion wheeleri, 173
Datura, 110; *D. inoxia*, 5; *D. stramonium*, 5, 38

Davis, Anthony, 59
delirium, 80–82, 103, 113, 123
Denber, Herman C. B., 116
depressant or stimulant substance, 205, 208, 232–36
Desert culture, 24
Digitalis, 112
3,4-dihydroxyphenylalanine, 149
2,5-dimethoxy-4-methylamphetamine, 128
3,4-dimethoxyphenylethylamine, 128, 143, 222
N,N-dimethyl-3-hydroxy-4,5-dimethoxy-phenethylamine, 142, 222
N,N-dimethyl-hydroxyphenethylamine, 115
N,N-dimethylmescaline, 128
DMM, 128
DMPE, 128, 143, 222
DOM, 128
dopa, 150–51
dopamine, 139, 150–51, 220

eagle claws cactus, 173
Echinocactinae, 167
Echinocactus, 155, 166–67;
    E. horizonthalonius, 173;
    E. jourdanianus, 216; *E. lewinii*, 213, 215, 219; *E. platyacanthus*, 173;
    E. pseudo-Lewinii hortulorum Thompsonii, 219; *E. rapa*, 212–13;
    E. texensis, 174; *E. williamsii*, 155, 212, 218–19; *E. williamsii* var. *anhaloninica*, 215; *E. williamsii*, "Hylaeid," α *pellotinica*, 218;
    E. williamsii, "Hylaeid," β *anhaloninica*, 215; *E. williamsii* var. *lutea*, 214; *E. williamsii* var. *pellotinica*, 218
Echinocereus: *E. pectinatus*, 173;
    E. reichenbachii, 173
Echinopsis (*Trichocereus*), 142;
    E. pachanoi, 21, 107, 143, 145;
    E. rhodotricha, 142; *E. terscheckii*, 143

Elk Hair, 26, 46
Ellis, Havelock, 92
El Santo de Jesus Peyotes, 11
ephedrine, 121
epinephrine, 115, 140, 149–50, 198
Epstein, Leon J., 184
Espostoa, 142; *E. huanucoensis*, 142
Euphorbia antisyphylitica, 173
"experimental analogs," 118
exquite, 118

Fabaceae, 24, 32, 140, 163
Fernberger, Samuel W., 87
Ferocactus: *F. echidnae*, 173;
    F. hamatacanthus, 173; *F. pilosus*, 173
Fiesta of the Peyote, 19–20
Firstborn Church of Christ, 46–47
"first fruits ceremonies," 34
fishhook or nipple cactus, 173–74
Flourensia cernua, 171, 173
Food, Drug, and Cosmetic Act, 197–98
formaldehyde, 140–41
Fouquieria: *F. formosa*, 171; *F. splendens*, 173
Frailea, 177
Frank Takes Gun, 196
Frederking, Walter, 116, 130
frijolillo, 33
Furst, Peter T., 13–14

Ganders, Fred R., 213
García, Bartholomé, 10
Ghost Dance, 24, 38–40, 44, 65
granjeno, 174
"Great Message" of the Iroquois, 40
Great Pyramid, 5
guayacan, 174
guayule, 174
Guttmann, Erich, 85–86, 99
Gymnocalycium, 142, 177; *G. gibbosum*, 143

hallucination, 84, 92, 97, 124, 132, 163
hallucinogen, 3, 118, 131, 198–200. See also psychotomimetic

hallucinogenic activity, 108; drugs and, 124, 197; effects, 92, 232; fungi and, 164–65; plants and, 3–4, 15, 22; substances and, 106, 116, 234
Harrison Narcotics Act of 1914, 185
Hayden, Carl, 191–92
Hechtia, 172; *H. glomerata*, 173, 175
Heffter, Arthur, 125, 138, 144, 154, 215
Hennings, Paul E., 138, 156–57, 213
Hensley, Albert, 27, 207–8
Hernández, Francisco, 6, 163
heroin, 134–35, 184, 186
híkuli, 14, 17, 19, 21–22
Hilliard, Judge, 196
Hoffer, A., 130
Hofmann, Albert, 92, 198
Hollister, Leo E., 119
Holmstedt, Bo, 144, 213, 218
hololisque, 7
hordenine, 115, 121, 133, 140, 142–44, 150–51, 220
Hordeum vulgare, 121
horse crippler, 174
H.R. 4230, 205
Huichol. See Native Americans of Mexico, Huichol
huisache, 174
human sacrifice, 5
Huxley, Aldous, 79, 86, 89–90, 102, 105
4-hydroxy-3-methoxyphenethylamine, 143, 221

index of aridity, 175
Inquisition (Spanish), 9, 188
insulin, 126
Ipomoea violacea, 165
isocitrimide lactone, 141, 230
isoquinolines, 139, 141, 224–28
ixtle, 16

Jatropha dioica, 173
Jáuregui O., Ernesto, 175
Jesus Christ, 29–30, 44–45, 53, 63–64, 66–67, 76, 208
jimson weed, 5, 38

Kapadia, Govind J., 139
Karwinskia humboldtiana, 174
Kauder, E., 138
Kazen, Judge, 196
Klüvier, Heinrich, 81, 99, 102, 104
Knab, Tim, 21
Knauer, Alwyn, 98, 100
Koeberlinia spinosa, 173
Koshiway, Jonathan, 46–47

La Barre, Weston, xiv, 3, 21, 41
Labiatae. See Lamiaceae
Lamiaceae, 165
Lancaster, Ben, 46
Landry, S. F., 111
Lanternari, Vittorio, 40
Larrea tridentata, 17, 169, 171–73, 175
Latter-Day Saints. See Church of Jesus Christ of Latter-Day Saints (Mormon)
Lawrence, Chris, 125–26
League for Spiritual Discovery, 197
leatherplant, 173
lechuguilla, 173–74
Leguminosae. See Fabaceae
Lemaire, Charles, 155, 212
Lennard, Henry L., 184
León, Nicolas de, 9
Leonard, Irving, 188
Leuchtenbergia principis, 173
Leucophyllum, 174
Lewin, Louis, 96, 138, 144, 153, 155, 157, 213–15
"Little Moon" ceremony, 46
Lophophora, 6, 162–68, 170, 175, 177–78, 220; *L. albilanata*, 219; botanical description of, 209–10; *L. caespitosa*, 218; common names of, 159–61, 210; *L. diffusa*, 143–44, 157–59, 163, 167, 171–72, 178, 180–81, 209, 216–18; distribution of, 210–11, 216–18; *L. echinata*, 214; *L. echinata* var. *diffusa*, 217; *L. echinata* var. *lutea*, 214; *L. flavilanata*, 219; *L. fricii*, 215–16; *L. jourdaniana*, 215–16; key to the

species of, 210; *L. lewinii*, 213;
L. lutea, 214; *L. texana*, 218;
L. tiegleri, 219; *L. williamsii*, front.,
xvii, 38, 143–44, 157–59, 166–70,
176, 178, 209–16, 218–19;
L. williamsii var. *caespitosa*, 218;
L. williamsii var. *decipiens*, 214;
L. williamsii var. *echinata*, 214;
L. williamsii var. *lewinii*, 213;
L. williamsii var. *lutea*, 214;
L. williamsii var. *pentagona*, 214;
L. williamsii var. *pluricostata*, 214;
L. williamsii var. *texana*, 218;
L. ziegleri, 218; *L. ziegleri* var.
albilanata, 219; *L. ziegleri* var.
diagonalis, 219; *L. ziegleri* var.
flavinata, 219; *L. ziegleri* var.
mammillaris, 219; *L. ziegleriana*, 218
lophophorine, 121, 138, 141, 144, 228
Loranthaceae, 140
LSD, 116–18, 124, 128–32, 164–65, 184, 198–200
lotebush, 173
Ludwig, Arnold M., 80, 83
Lumholtz, Carl, 13–14, 21
lysergic acid alkaloids, 165
lysergic acid diethylamide. *See* LSD

Maclay, W. S., 85–86
Mahonia trifoliata, 19
maize-roasting ceremony, 20
maleimide, 141, 229
malimide, 141, 229
Maloney, William J.M.A., 98, 100
Mammillaria, 142, 157, 166–67, 173–74;
M. cirrhifera, 219; *M. compressa*, 219;
M. lewinii, 213; *M. longimamma*,
162; *M. pectinifera*, 142, 162;
M. williamsii, 212
marijuana, 112, 129–32, 137, 165, 184,
187, 198–200, 234
Matorral Xerófilo, 171, 174
McAllester, David P., 65
McCleary, James, 115
McFate, Yale (judge), 195

McKellar, Peter, 88
McLaughlin, Jerry, 115, 139, 150
Merlis, Sidney, 116
mescal, 163
mescal bean, 32, 34, 38, 163; ceremony, 163; medicine society, 34; society, 33; visions produced by, 33
mescal buttons, 113
mescaline, 79, 81–83, 92, 107, 111, 116–17, 143–44, 198–200, 223, 234; biosynthesis of, 150–51; chemical structure of, 132–34, 137–41, 144–49; clinical use of, 115–19; compared with LSD, psilocybin, and marijuana, 129–32; cytogenetic effects of, 120, 128–29; dosages of, 120, 125; effects of, compared to schizophrenia, 118–19; extraction of, 146–47; hydrochloride, 125, 145; physiological effects of, 122–25; structure-activity relationship of, 147–49; sulfate, 125, 145; synthesis of, 146–47; tolerance to, 127–28; toxicity of, 120, 125–27. *See also* Peyote
mescalism, 31–32
mescaloruvic acid, 141, 230
mescalotam, 141, 230
mescaloxylic acid, 141, 230
"mescal psychosis," 81
mesolithic, 3
mesquite, 169, 174–75
3-methoxy-4,5-methylenedioxy (myristicin), 137, 198
O-methylanhalonidine, 138, 226
N-methylmescaline, 138, 142, 223
N-methyltyramine, 142, 150, 220
mezquital, 174
Microphyllous Desert Scrub, 171
Mimosa, 173
Mitchell, S. Weir, 93–94, 101
mitosis, 129
"model psychoses," 119
Mooney, James, 42–43
Morgan, F. P., 113
Morgan, G. R., 23

Mormons. *See* Church of Jesus Christ of Latter-Day Saints
morning glory, 4
morphine, 113, 134–35
Mortimer, Raymond, 82
Muller, Cornelius, 172
Muller, K., 14–15
Murphree, Henry B., 198
"muscale button," 155, 213
mushrooms, 4
Myerhoff, Barbara G., 13
Myristica fragrans, 198
myristicin, 137, 198
Myrtillocactus geometrizans, 169, 178

narcotic, 5, 107, 111, 127, 183, 195, 200, 232–36; definition of, 185–87
Native American Church, xiii, 46–49, 51–53, 55, 59, 63–64, 79, 111, 180, 190, 196–97, 199–202, 204–5, 207; of Canada, 47; of Navajoland, 48; of North America, 48, 196; of Oklahoma, 48; of South Dakota, 48
Native Americans of Mexico, tribes: Chichimeca, 6–7, 159; Cora, 11, 13, 23, 160; Huichol, xvii, 13, 14–21, 50, 128, 160; Nahuatl, 5–6, 160; Tamaulipan, 32; Tarahumara, 13, 21–23, 32, 43, 109–10, 160, 169; Tepecano, 13; Yaqui, 108
Native Americans of the United States: agriculture of, 36; attitude of federal government toward, 34–37; changes in way of life of, 35–38; discrimination against, 34–38; disease among, 36–37; fasting by, 37; interaction of, with Anglos, 34–42; loss of pride of, 36; peyotism and, 39–48; religious beliefs of, 37–48; self-torture among, 37–38, 41; vision quest of, 33, 37–38, 40–41
Native Americans of the United States, tribes: Apache, 30–31, 33, 59; Arapaho, 44, 53, 59, 64; Caddo, 33, 44; Carrizo, 43, 58; Carrizo Apache, 30; Cheyenne, 44, 53, 59;

Comanche, 30–34, 43–45, 50, 53, 107, 160; Dakota, 37; Delaware, 26, 33, 37, 39, 44, 46, 64, 107, 160; Gosiute, 61; Iowa, 33, 59, 64; Iroquois, 40; Kansa, 33; Karisu: *see* Carrizo; Kickapoo, 107, 160; Kiowa, 31–32, 42–43, 67, 70–76, 160; Kiowa Apache, 107; Kiowa Comanche, 54; Laguna, 67; Lipan Apache, 31–32, 43, 53, 58; Menomini, 28, 30, 63; Mescalero Apache, 27–28, 31, 43, 60, 160; Navajo, 42, 52, 61, 66–77, 107, 160, 193, 195–96; Northern Paiute, 38, 44, 59, 61; Omaha, 33, 46, 107, 160; Opata, 160; Osage, 33, 193; Oto, 33–34, 43; Otomi, 160; Pawnee, 33, 59, 61; Ponca, 33; Shawnee, 107; Sioux, 38–39; Tepehuane, 160; Taos, 107, 160; Tonkawa, 33–34, 59; Ute, 63; Washo, 59, 61; Wichita, 33, 160; Winnebago, 27, 37, 46, 52, 60, 64–65, 103, 108, 160, 207
Navajo Tribal Council, 194
Navajo V-way ceremony, 66–77
Neo-American Church, 197
Neolloydia conoidea, 173
Nettl, Bruno, 65
Nicotiana: *N. rustica*, 16, 164; *N. tabacum*, 4, 38
nicotine, 135, 164
norepinephrine, 124
nor-mescaline, 150
nutmeg, 137, 198

Obregonia, 142, 166–68, 178; *O. denegrii*, 162
Ochoterena, I., 213
ocotillo, 171, 173
Oklahoma Enabling Act, 189
old peyote complex, 23, 25, 31–32, 43
ololiuqui, 10, 20, 164–65
opiates, 200, 233
opioids and opioid antagonists, 186
opium, 109, 113, 127, 134, 187, 199–200, 233
Opuntia, 6, 142; *O. engelmannii*, 169,

174, 181; *O. leptocaulis*, 174;
  O. stenopetala, 174; *O. tunicata*, 174
Opuntioideae, 177
Oregon: Court of Appeals, 202;
  Supreme Court, 202–3
Ortega, José, 11
Osmond, Humphrey, 86, 118, 130, 193

Pachycereus, 142; *P. pecten-aboriginum*, 143
Paleo-Siberian, 4
Palmer, Edward, 109
Panaeolus, 164; *P. campanulatus* var. *sphinctrinus*, 164
Pan-Native American movements, 38, 40–41, 44, 47–48
Papaver somniferum, 134
Parke, Davis, and Company, 138, 153, 155, 213–14
Parker, Quanah, 45, 54, 108
Paul, A. G., 115
Parthenium incanum, 174
patent medicines, 185
"peiotl," 159
Pelecyphora, 142, 146, 166–67;
  P. aselliformis, 142–43, 145, 162;
  medical use of, 145
pellotine (pellotin), 121, 138, 141, 143–44, 151–52, 227
pencil cholla, 174
Pereskia, 177
Pereskioideae, 177
Pereskiopsis velutina, 178–79
Petrullo, Vincenzo, 39, 89
peyocactin, 115
peyoglunal, 141, 231
peyoglutam, 141, 230
peyonine, 141, 231
peyophorine, 141, 228
peyoruvic acid, 141, 152, 231
"peyot," 7, 13, 160
peyote, front., 159–65, 184–87, 206, 234;
  alkaloids of, 138–41, 143–52; alkaloids of, in other cacti, 141–43; amino acids of, 141; antibiotic action of, 114–15; biogeography of, 168–71;
biosynthesis of alkaloids of, 149–52;
bird, 72, 75; botanical history of, 155–59; "buttons" of, 50–51, 56, 70–71, 109, 125, 156, 179–80; chemical differences between species of, 143–44, 213–14; chromosome number of, 166; classified as a narcotic, 186; common names of, xvii, 11, 159–61; conservation of, 180–82; cultivation of, 178–80, 207; cytogenetic effects of, 128–29; ecology of, 171–75; evolution of, 177–78; first illustration of, 156, 212; gardens, 51–52, 59, 181; legal classification of, 185–87, 197–201, 234; "magical powers" of, 41; medical use of, by Native Americans, 4–7, 11, 13–14, 22, 29, 44–45, 49, 106–11, 114, 187; medical use of, by Anglos, 111–14, 119; morphology of, 165–68; origin of the word, 159–60; pharmacology of, 120–32; physiological effects of, 120–25; physiological effects of alkaloids of, 120–25; plants confused with, 162–65; populations of, 175–77; psychiatric uses of, 115–19; reproductive structures of, 166–67, 176; as a sacrament, 63, 78–79, 190, 201; songs used in ceremonies of, 56–58, 65–66; Spanish Conquest and, 4–11, 24; spelling variations of, 160; as a Spirit, 25, 27–28, 41, 44–45, 64; use by Native Americans in Mexico, 11–24; vegetative structure of, 165–66. *See also* Mescaline; Navajo V-way ceremony; peyotism
peyote, experience caused by use of, 79–105; absence of dreams, 84, 91; abnormal emotional states, 81; abnormal sensory phenomena, 81; adverse reactions, 102–4; alteration in conscious states and attitudes, 80–81, 84; depersonalization, 79, 83–84; difficulty of communication, 84, 87; distortion of time and space, 84, 86–87; dual existence, 84–86; effects on

peyote (*continued*)
  memory and thinking, 82, 84, 88; emotional extremes, 84, 89–91; euphoria-anxiety relationship, 82; hallucinations, 84, 92, 97; inhibition of sex drive, 84, 91; mixing and heightening of the senses, 84, 88–89; modification, 81; phases, 83–84; physiological changes, 83; problems in understanding and describing, 80, 87; significance of setting, 81; synesthesia, 84, 88–89
peyote, visual experience caused by, 92–102; after-images, 98; changes in brightness and saturation, 99; color discrimination, 90, 100–101; constant change of vision content, 101–2; contrasts, 99; dimensionality, 102; general visual effects, 98–102; halos, 98; movement, 99; size and symmetry, 99; stages, 96–98
peyoteros, 50–51, 59, 181, 207
"peyote tribe." *See* Native Americans of Mexico, Huichol
"peyotillo," 145, 162
peyotism: anti-Christian emphasis of, 46; attraction of, 39–42; basic plains ceremony of, 54–58; biblical validity of, 63; Christian elements in, 40, 54, 62–64, 76–77; Christian influences on, 28, 40, 43, 46, 62–64; and "Father Peyote" or "Chief Peyote," 49, 55, 57–58, 61–62, 69, 73; moon rites of, 53; music of, 56–58, 65–66, 70–71, 73–75; old peyote complex of Mexico, 23–25, 31–32, 43; origin of, 23–25; paraphernalia, 25, 53, 55–56, 63, 67, 69, 75; and the peyote "road" or "way," 41–42, 44–45, 53, 63; and pilgrimages for peyote, 50; and preparations for ceremony, 52–54; prophets of, 44–46; and smoking, 47, 56, 60, 69–71; spread of, 42–49; types of peyote and their uses in, 58–59; variations in the ceremony of, 58–62; and visions, 31, 40–41, 62, 77; and the V-way ceremony, 66–77
Péyotl, 6, 10, 160; *P. Xochimilcensi*, 6, 163; *P. Zacatecensi*, 6, 163
peyoxilic acid, 141, 231
Pfeiffer, Carl C., 198
phenacetine, 113
β-phenylethylamine, 133, 139–40
phenol, 139
phenolic compound, 139
phenylalanine, 140, 149–51
phenylethylamine, 139–40, 144, 149–51, 198, 220–24
Philip II (king of Spain), 6
photosynthetic active radiation (PAR), 180
physostigmine, 126
piciete, 7–8
picietl, 164
piperine, 134
pipiltzintzinli, 165
Plains peyotism. *See* peyotism
Polaskia, 142
Pomona College, xiii
Porlieria angustifolia, 174
poyomate, 7
Prentiss, D. W., 113
prickly-pear cactus, 6, 169, 174
Prieto, A., 13
Prosopis glandulosa, 169, 171, 174–75
"pseudo-hallucination," 92
psilocin, 165, 234
Psilocybe, 5, 164
psilocybin, 118, 128–32, 165, 198–200, 234
psychedelic therapy, 117
psychoactive substances, 124; plants, 4, 11, 38, 164; properties of, 119, 134
psycholysis, 117
psychotomimetic substances, 80–81, 92, 115–20, 123–24, 127, 131, 136–37, 139, 198
Public Laws: 91–513: 232–36; 95–341: 201, 236–37; 103–41: 237–38; 103–344: 205, 238–39

quinine, 135

Radin, Paul, 103, 108
Rain, Howard, 28
Ransom, Donald C., 184
Rao, G. Subba, 115
rainbow cactus, 173
rarikira, 20
Rave, John, 45, 59, 62, 64, 103, 108
religious freedom, 187–97, 201–5
Restoration Act, 204, 237–38
retama, 174
Rhodes, Willard, 65
Rio Grande Plains, 175, 206
Rivea corymbosa, 4, 160, 164–65
Rosettophyllous Desert Scrub, 171–72
Rouhier, Alexandre, xiv
Rümpler, Theodore, 155
Rzedowski, Jerzy, 171

"sacred mushroom," 164
Safford, William E., 164
sagebrush, 55–56, 69
Sahagún, Bernardino de, 5–6, 159
Salm-Dyck, Prince, 155, 212–13
Salvia divinorum, 165
Sands, David, 145
San Pedro cactus, 21, 107, 145
Santa Niña de Peyotes, 11
Schaefer, Stacy B., 13–14
schizophrenia, 118–19
Schultes, Richard Evans, 92, 106–7, 110, 198
Seevers, M. H., 145, 186
Senecio, 163; *S. hartwegii*, 160
serotonin, 123–24
Shakerism, 40
shaman, 4, 14–19, 21–23, 37, 110
Shonle, Ruth, 62
Simpson, Lorna, 88
Sinnett, E. Robert, 85
Slotkin, J. S., 23, 32
small-dose technique, 117
Smet, P.A.G.M. de, 21
Smith, Al, 202

Smythies, John, 118
Solanaceae, 163–64
Sophora secundiflora, 24, 32, 38, 163
sotol, 173
Soto Mora, C., 175
South Texas Plains, 174
Spaniards, 4–7, 9–11, 13, 21, 24, 43, 163–64, 188
Späth, Ernst, 138, 146
Speck, Louise B., 126
Staphylococcus aureus, 115
Stenocereus, 142
Stetsonia, 142; *S. coryne*, 142–43
Stewart, Omer C., xiv, 23
Strombocactus, 168, 178
S. disciformis, 163
Stropharia, 164; *S. cubensis*, 165
strychnine, 112, 121, 135
Strychnos, 112
succinimide, 141, 229
sympathomimetic agents, 123
Szara, Stephen, 198

Tamaulipa: brushlands in, 174; thorn forest in, 168, 173
Tamaulipeca, 174
tarbush, 173
"Technicolor hallucinations," 98
"teonanacatl," 164
tesvino, 19, 23
tetrahydrocannabinol. *See* THC
tetrahydroisoquinoline, 149–52, 224–28
THC, 131, 137, 198, 234
Thelocactus, 166–67
T. setispinus, 174
Thevetia thevetioides, 5
Thomas, Elmer (senator), 193
tlitlitzen, 165
tobacco, 3–4, 16, 32, 38, 50, 55, 60, 69, 184
Tobriner, Judge, 196
Todd, James S., 143, 213
tolerance, 127–28, 131, 185–87; reverse, 131
toluene, 139

Index : 271

"toxic delirium," 82–83
Trans-Pecos Region, 24
3,4,5-trimethoxyamphetamine, 148
3,4,5-trimethoxybenzoic acid, 124
3,4,5-trimethoxy-β-phenethylamine. *See* Mescaline
3,4,5-trimethoxyphenylacetic acid (TMPA), 124
3,4,5-trimethoxyphenylisopropylamine (TMA), 148
Troike, Rudolph, 33
tryptamine, 140, 198
tryptophan, 137
Turbina corymbosa, 20
Turbinicarpus, 142, 168; *T. pseudopectinatus*, 163
tyramine, 140, 142, 150, 220
tyrosine, 140, 149–51

Underhill, Ruth, 31, 65, 125
Union Church, 47
U.S. Congress, 201; actions since 1990, 204–5
U.S. Constitution, 187; First Amendment to, 188, 190, 196, 202, 204; Fourteenth Amendment to, 195
U.S. Supreme Court, 188, 195; decision of 1990, 201–5
'uxa, 19

Viscum album, 140

visual experience. *See* Peyote, visual experience caused by

Wagner, Roland M., 61
Wallace, Robert, 168
Warnock, Barton, 168–69
Wasson, R. Gordon, 160
Watson, Sereno, 214
waxplant, 173
White Thunder, 59
Wilson, Jack, 38–39, 44
Wilson, John, 44–46, 62, 64
Wirikuta, xvii, 15, 17–19
withdrawal syndrome, 187
Wodziwob, 38
woqui, 46
Wounded Knee, 39
Wovoka, 38–39, 44

Yellow Bird, 39
yoyotli, 5
Ysla, D. Pedro Nabarre de, 188
Yucca, xvii, 169, 172; *Y. carnerosana*, 174; *Y. elephantipes*, 175; *Y. filifera*, 169, 174–75

Zaehner, R. C., 85, 91, 100, 104
Zingg, Robert M., 22, 109
Ziziphus obtusifolia, 175
Zumarraga, Juan de, 5

ABOUT THE AUTHOR

Edward F. Anderson's interest in cacti and tropical plants was originally sparked by a fellowship to study various Mexican cacti, including peyote, at the Rancho Santa Ana Botanic Garden in Claremont, California. After earning a B.A. in biology from Pomona College and an M.A. and a Ph.D. in botany from the Claremont Graduate School, Anderson taught botany at Whitman College in Walla Walla, Washington, for thirty years, during which time he received two Fulbright-Hays Lectureships to teach overseas. He also spent sabbatical leaves studying cacti and tropical plants in Latin America and Southeast Asia.

Author of more than twenty scientific papers on conservation, ethnobotany, and taxonomy, Anderson is also the author of the book *Plants and People of the Golden Triangle* and coauthor of *Threatened Cacti of Mexico*. He has been honored by several cactus and succulent societies and was elected to membership in the prestigious Linnean Society of London. He also served for six years as president of the International Organization for Succulent Plant Study. Anderson now serves as an ethnobotanist on the science advisory team of Shaman Pharmaceuticals, and he is Senior Research Botanist at the Desert Botanical Garden in Phoenix, Arizona.